PUBLISHED FOR THE MALONE SOCIETY BY
OXFORD UNIVERSITY PRESS

WALTON STREET, OXFORD OX2 6DP

Oxford New York Toronto
Delhi Bombay Calcutta Karachi
Kuala Lumpur Singapore Hong Kong Tokyo
Nairobi Dar es Salaam Cape Town
Melbourne Auckland Madrid

and associated companies in
Berlin Ibadan

ISBN 0 19 729031 0

Printed by BAS Printers Limited, Over Wallop, Hampshire

COLLECTIONS
VOLUME XV

THE MALONE SOCIETY
1993

CONTENTS

RALPH CRANE'S TRANSCRIPT OF *A GAME AT CHESS*, BODLEIAN MANUSCRIPT MALONE 25

This edition of *A Game at Chess*, from Manuscript Malone 25 in the Bodleian Library, was prepared by the General Editor and checked by Katherine Duncan-Jones.

November 1992 N. W. BAWCUTT

INTRODUCTION

THE manuscript reprinted here, Bodleian MS. Malone 25, is one of three transcripts made by Ralph Crane of Middleton's *A Game at Chess*, the other two being British Library MS. Lansdowne 690 and the Archdall manuscript, pressmark V.a.231, in the Folger Shakespeare Library, Washington. All three are in the quarto format used by Crane in making presentation copies for private patrons: Malone's pages are 7.25 inches deep by 5.75 inches wide, Lansdowne's 7 by 5.5, and Archdall's, which have obviously been severely cropped during the original binding, 6.5 by 4.9. All have marginal rules in ink, forming a box that encloses the text of the play, with catchwords in the bottom right-hand corner. Page numbers are provided at the top of the page; Archdall alone has running titles. They are all written in the elaborately calligraphic manner customary with Crane, though Archdall makes a noticeably less frequent use of italic words and brackets than the other two.

The three texts bear no sign whatever of theatrical use, and no clear indication that any of them was derived from prompt-copy, but they are by no means identical. Archdall bears on its title-page the date '13 August 1624', near the end of the play's brief run (5–14 August); it is, however, clearly derived from an early version of the play which Middleton revised before production. During this revision Middleton introduced the character of the Fat Bishop, adding substantial sections concerning him in II. ii and III. i, and in Act IV transferring to him lines formerly spoken by the Black Bishop, who from then on has a very minor role. Middleton also added lines (I. i. 348–53 and III. i. 303–5 in R. C. Bald's edition of the play (Cambridge, 1929); 454–9 and 1371–3 in the society's edition of the Trinity manuscript, edited by T. H. Howard-Hill, 1990)[1] which identified the treacherous White King's Pawn more clearly as Lionel Cranfield, Earl of Middlesex, who fell from power as Lord Treasurer in May 1624.

Malone and Lansdowne, on the other hand, are late texts which exhibit a number of errors that crept in during Crane's repeated copyings of the play (he must certainly have prepared other copies which are now lost, in addition to the three which survive). Lansdowne cannot be precisely dated, but according to his dedication Middleton intended Malone to be presented to William Hammond on New Year's Day, 1 January 1625, so it was presumably completed

[1] In the rest of the Introduction, references to Trinity will be given twice: act, scene, and line references will be to Bald's edition, through-line references to Howard-Hill's. Another helpful edition of the Trinity manuscript is Susan Zimmerman Nascimento's Ph.D. dissertation, 'Thomas Middleton's "A Game at Chesse": A Textual Study' (University of Maryland, 1975), containing detailed collations of the manuscripts and printed editions.

in the last few weeks of 1624. The Malone manuscript is of particular interest to Shakespeare scholars because it exhibits the 'massed entries' (of which more will be said later) found also in three First Folio comedies, *The Two Gentlemen of Verona*, *The Merry Wives of Windsor*, and *The Winter's Tale*, which are generally thought to have been printed from transcripts by Crane. It has also been drastically abridged. At the same time it has striking affinities with the Lansdowne version, and a comparison of the two helps to throw light on the distinctive features of Malone. It would be impossible to reprint all three of Crane's versions of the play, and the procedure adopted here has been to reprint Malone in its entirety, because of its unusual characteristics and its relevance to the First Folio, followed by selected pages from Lansdowne for purposes of comparison. (From now on these manuscripts will be referred to as 'L' and 'M'.)

M survives in what may well have been its original presentation binding of parchment with gilt stamps and holes for tie-strings or ribbons which are now missing. The binding has come away from the spine, thus aiding the determination of the gatherings. Crane himself numbered the pages of the play (i.e. excluding the preliminaries and induction) from 1 to 69; the opening section has been given roman numeration in pencil up to page xii by a later hand, most probably that of a nineteenth-century librarian of the Bodleian, who also numbered the final blank verso '70(ult.)'. When this numeration was made the manuscript began with a two-leaf gathering of blank paper, which seems to have been present when R. C. Bald examined the manuscript during the preparation of his edition. The outermost leaf has disappeared, and the manuscript now begins with page iii, bearing the signature 'J Pepys' and library stamps; page [iv] is blank.

The next section, which will be called 'B' for ease of reference, consists of five leaves containing the following material: page v, title-page in box-rules, decorated with corkscrew curls; page [vi], blank except for box-rules; page vii, Middleton's autograph dedication of the manuscript to William Hammond; page [viii], blank; pages ix–xii, two leaves containing the induction; pages 1–2, the first leaf of the play itself. Clearly the dedication leaf was prepared separately by Middleton himself and tipped-in during the process of binding. Crane's ink was in varying shades of brown, whereas Middleton's was pale, greyish black, and Middleton's box-rules, in the same ink as his dedication, are not as tidy and symmetrical as Crane's. Then follow eight unsigned gatherings each of four leaves, to which signatures will be assigned as follows: C, pages 3–10; D, pages 11–18; E, pages 19–26; F, pages 27–34; G, pages 35–42; H, pages 43–50; I, pages 51–8; J, pages 59–66. The manuscript ends with a two-leaf gathering (K) consisting of pages 67–9 and a blank verso to page 69.

All the watermarks present are of an elaborate left-handled pot-shape with a crescent similar to number 3627 in Heawood's *Watermarks mainly of the 17th and 18th Centuries* (Hilversum, 1950). There are, however, variations in the markings on the belly of the pot. The commonest form, consisting of the letter G above the letters RO, occurs in gatherings B, C, D, E, G, and K. In gatherings F, H, and I, the markings consist of the letter O, with a slanting line over it rather like a French grave accent, above the letters RO. Gathering B has two watermarks. In addition to the one already mentioned, the dedication leaf exhibits the bottom half of a pot similar in outline to the other watermarks, but with what appears to be the head of a dog or fox, with open jaws pointed towards the base of the pot, on the belly of the pot. No watermark can be found in J, or on the isolated leaf which forms pages iii–iv.

A plausible identification of the dedicatee William Hammond has been made by R. C. Bald and Margot Heinemann.[2] As Bald noted in his edition (p. 137):

> Mr William Hammond seems to have belonged to a coterie of Middleton's civic patrons. Middleton's pious compilation, *The Marriage of the Old and New Testament* (1620) is dedicated to 'the two Noble Examples of Friendship and Brotherhood, Mr. Richard Fishborne, and Mr. John Browne.' Fishborne (a Mercer) and Browne (a Merchant-Tailor) were partners; and in Richard Fishborne's will Browne is named as the executor, while William Hammond was one of twenty friends who were each left five pounds for a mourning ring.

Fishborne's will was dated 30 March 1625 and proved on 14 May of the same year.[3] Another recipient of a mourning ring was 'Mr Edmond Hamond', and in an annexe to the will (fol. 464) 'Mr Edward Hamond' [*sic*, presumably a scribal error] and 'Mr William Hamond' were each left five yards of mourning cloth at a value of seven pounds and ten shillings, and a cloak worth forty shillings for his manservant.

The two Hammonds seem to have been brothers. Edmond Hammond lived in the London parish of Allhallows on the Wall, and a memorial tablet, still surviving in the church, records that he died on 24 April 1642 at the age of 67. His will, proved on 28 April 1642, made elaborate provisions for a large number of family and charitable bequests.[4] 'Mr. Wm. Hammond' was buried at Allhallows on 10 January 1634/5.[5] He apparently died intestate, and on 16 January 1634/5 in the Archdeaconry Court of London administration of his estate was granted to his brother Edmond ('Edmundo Hammond fratri'), as

[2] *Puritanism and Theatre: Thomas Middleton and Opposition Drama under the Early Stuarts* (Cambridge, 1980), 169–70.

[3] PRO PROB 11/145, fol. 461 ff.

[4] PRO PROB 11/189, two apparently identical copies numbered 44 and 49.

[5] E. B. Jupp and R. Hovenden (eds.), *The Registers of . . . the Parish of Allhallows London Wall 1559–1675* (London, 1878), p. 281.

recorded in the Probate and Administration Act Book no. 6 for 1626–37.[6] An inventory of his estate shown to the court valued it at £3,704; by the standards of the time he was a wealthy man. Edmond Hammond was admitted to the Haberdashers' Company on 21 July 1598,[7] but William does not appear to have been a member. The documents make no mention of a wife or child for Edmond or William, and both men were probably bachelors.

M bears the signature 'J Pepys' on the recto of the leaf preceding the title-page. There were several bearers of this name in the seventeenth century, including the father of the diarist. F. P. Wilson suggested that he might be the John Pepys (1587–1642) of Cottenham, Cambridge.[8] But as Wilson himself noted, this John Pepys was a disreputable character who was cited in the ecclesiastical courts, probably in 1620, for loose living and failing to maintain his wife,[9] and his will shows that he died heavily in debt.[10] It is surely more probable that the signature belongs to the John Pepys (1576–1652) of Salisbury Court, London, who is mentioned in Pepys's diary on 30 January and 3 December 1668; he had a long association with Lord Chief Justice Coke, and became rich enough to run his own coach.[11]

For two centuries the manuscript vanished from sight, until in the mid-nineteenth century it was put on sale in a catalogue issued by a London bookseller, C. J. Stewart. In the introduction to his text of *A Game at Chess* A. H. Bullen mentioned the catalogue as having appeared 'some twenty years ago', and regretted that he had never seen the manuscript: 'I have hunted for this MS. high and low, but I have not been able to discover who is the present possessor of it.'[12] No copy of Stewart's catalogue has yet been traced, but the entry on Middleton in Joseph Hunter's *Chorus Vatum Anglicanorum*[13] establishes that the catalogue appeared in 1858 and that the printed slip gummed into the inside front cover of M is a cutting from it. Stewart argued from the manuscript's brevity that it was a first draft, and conjectured that William Hammond was the 'Mr. W.H.' of Shakespeare's *Sonnets*. Hunter was rightly sceptical about the latter point, and modern textual scholarship would not accept

[6] Guildhall Library MS. 9050/6, fol. 159^{v}.

[7] Membership List of the Haberdashers' Company 1526–1641, Guildhall Library MS. 15857, fol. 143^{v}.

[8] 'Ralph Crane, Scrivener to the King's Players', *The Library*, 4th ser. vii (1926–7), 210.

[9] See Bodleian MS. Tanner 167, fols. 72–3.

[10] PRO PROB 11/189/57.

[11] For more information on the two men see W. H. Whitear, *More Pepysiana* (London, 1927), pp. 29–32 and 42, and Edwin Chappell, *Eight Generations of the Pepys Family* (London, 1936), pp. 36–8.

[12] *The Works of Thomas Middleton*, 8 vols. (London, 1885–6), vii, 3.

[13] Vol. i. p. 60 (BL MS Add. 24487).

the former. M was acquired by the Bodleian Library, but the precise date of purchase cannot be established. The cataloguing of the work as part of the Malone collection is perhaps misleading: there is no evidence that Edmond Malone ever owned it or even knew of its existence.

M is written in Crane's mixed hand, and as with his other play transcripts it is often very difficult to tell whether or not a particular word should be printed in italic. The principle followed here has been to print italic only where the general appearance and size of the word seemed to indicate fairly clearly that italic was intended. It is also frequently hard to tell whether or not capitalization is intended, particularly with the following letters: C D F H L M N O W Y. Crane's *e* is sometimes so vestigial that it could be read as *i*; occasionally at the end of a word it is enlarged and could be mistaken for *d*. When a word ends in *t* the cross-bar may be omitted (see notes on 1052, 1192, 1200, and 1578); at 363 what appears to be a hyphen may be an extended cross-bar of the preceding *t*. With words containing *-ing* Crane once or twice links together the first two minims, and places the dot for *i* over *g*, so that it appears to read *-nig* (for examples see notes at 752 and 1479).

Crane's heavy punctuation, involving frequent use of hyphens, colons, and brackets, is well illustrated in M. Double hyphens can be found at 689 and after the catchword on page 53. Question-marks where the lower element has been enlarged to a comma can be found six times, at 100, 332, 1100, 1164, 1174, and 1398; more elaborate combinations of question-marks and commas occur at 1218 (compare Trinity *Game*, 2194), and 1383. Pen- or pencil-strokes through words or lines are noted at 1096, 1278, and 1602–5; these are unobtrusively written and do not appear to be deletions or corrections by Crane. Their presence is hard to explain, as is the existence throughout the manuscript of numerous short horizontal pencil-strokes or dashes, normally about an eighth of an inch long and placed between or at the end of the lines of text. Markings of this kind occur on or immediately below the following lines (an asterisk indicates that a dash is placed at the end of the line and could be mistaken for genuine punctuation): 13, 17, 25, 35, 38, 46, 47, 52, *61, 75, 78, 103, 114, 127, 146, 192, 207, 214, 217, 221, 222, 230, 233, 259, 288, 291, 292, 297, 298, 310, 323, 324, 332, 342, 355, 363, 375, 377, 379, 388, 425, 428, 452, 490, 493, *499, 504, 515, 526, 528, 534, *539, 546, 549, 565, 573, 605, 623, 627, 634, 644, 652, 669, 673, 705, 719, *732, 737, 782, 794, 860, 892, 919, 929, 939, 946, 974, 978, 985, 1027, 1042, 1049, 1052, *1058, *1063, 1065 (twice), 1066, *1067, *1072, 1076, 1084, 1089, 1094, 1096, 1104, 1110, *1115, 1123, 1134, 1136, 1140, 1163, 1168, *1171, 1172, 1180, 1182, 1184, 1195, 1199, 1200, 1207, 1208, 1210, 1214, 1221, 1228, 1234, 1239, 1248, 1251, 1254, 1255 (twice), 1260, 1262, 1273, 1275, 1285, 1287, 1295, 1296, 1297, 1304, 1316, 1335, 1364, *1378, 1398, 1403, 1412,

1428, *1430, 1434, *1445 (twice), 1467, 1469, 1472, 1473, 1477, 1478, *1483, 1485, 1501, 1506, 1511, *1523, 1539, 1542, 1548, 1553, 1566, 1574, 1576, 1605, 1606, 1611, *1633, 1635 (twice), 1643, and 1645.

It is unlikely that M is derived directly from L, since M contains two lines (529–30) and one phrase ('How doe you?', 935) which were omitted from L. In addition, M includes, perhaps rather surprisingly, three stage-directions, of which details will be given later, not found in L. L and M also have divergent readings at several points. At the same time, there is obviously a close relationship between the two texts, and they both contain a number of distinctive features found in no other manuscript or printed copy. One is the complete omission of the scene between two black pawns and a white pawn (III. ii in Bald's edition) and the running-together of the two scenes surrounding it. Another is the omission of a line (I. i. 290; 395), followed by a mistaken 'correction' in the next line. The Trinity version of the passage reads as follows:

> Hee spyes mee now, I must uphold his reuerence
> espetiallie in publick, though I knowe
> Priapus, Guardian of the Cherrie Gardens
> Bacchus and Venus Chit, is not more Vitious
>
> ———
>
> Bl. Bs. p. Blessings Accumulation keepe with you (Sir,)
>
> (I. i. 288–92; 394–8)

L reads:

> He spies me now: I must vphold his *Reuerence*
> (especially, in publique) though I know
>
> [bottom of page 17 of ms: catchword *Bacchus*]
>
> *Bacchus*, and *Venus Chitt*, are not more Vitious.
> *Bl. B*ˢ. *P*. Blessings Accumulation keepe with you (Sir)

The corresponding passage in M, lines 308–11, repeats the omission and false correction ('are' for 'is' in line 310), but adds two more errors of its own, 'privat' for 'publique' in line 309, and 'be' for 'keepe' in line 311.

Another unique omission in L and M is much neater, and it is hard to tell how deliberately it was done. Some lines in Trinity read:

> but tis in our power now
> to bring time neerer, Knowledge is a Mastrie,
> and make it obserue us, and not wee it:
>
> (IV. i. 52–4; 1659–61)

L and M omit the second line and modify the third:

> but 'tis in our powre now
> to make it obserue vs, and not we it.
>
> (968–9)

In one instance only the unique feature is an addition, but the two texts treat it in different ways. At IV. i. 41–3; 1648–50 Trinity reads:

> wh. Q^s. p. oh my heart!
> Bl. Q^s. p. that tis!
> wh. Q^s. p. the uerie selfe same . . .

M at 958–9 adds a phrase found elsewhere only in L, but deletes the second line of Trinity:

> *wh. Q^s. P.* oh, my hart: 'tis He:
> the verie self-same . . .

L is the only text of the play to have both phrases:

> *wh. Q^s. P.* oh, my hart: 'tis *he*:
> *Bl. Q^s. P.* that 'tis.
> *wh. Q^s. P.* the very self-same . . .

It would certainly seem that 'that 'tis' is a better reply to ''tis he' than to 'oh my heart', and there is a possibility that L is uniquely correct at this point.

The remaining variants exclusive to L and M are more straightforward, and may simply be listed. (The reading of the Trinity manuscript is on the left, that of L and M on the right.)

100	they're	Theis are
168	Vertue that	Vertue
257	fingers, by this hand	Fingers
330	the	your
400	Vertues	Vertue
444	life sir,	life
484	hah? by this hand	*Anglica*:
553	a	on
555	prouided	prescribd
556	inuented	provided
564	oh Insufferable!	Insufferable:
Between 598 and 599	wh. Q. hah?	*omitted*

609	cursed	haples
658	why well	well
667	while, I ha' gapte fort	while?
683	Noble	*holy*
689	submisse	*submissiue*
730	a whistle or a whisper	a Whisper, or a Whistle
777	that	this
794	his Attempt	him
854	honors	*Honor*
888	upward	vpwards
1051	you	'em
1139	thirtie thre	*Thirteene*
1501	plumpe	fat

Most of these readings are omissions or errors, and none appears to have any authority.

The numerous textual divergences between L and M fall into four groups. The first consists of accidentals. A comparison of the two inductions, for example, shows that though substantive readings correspond there are many differences in punctuation and in the use of italics, hyphens, and capitals. There are a few variants in spelling-forms (e.g. line 21, noe/no and line 47, readie/ready) and in contractions (e.g. lines 53 and 81, It's/'tis and line 82, 'em/them).

The second group contains variants produced by revision of M during the process of abbreviation. In the following list L, in the left-hand column, agrees with Trinity. (Line 1247 is not in Trinity, which omits the passage corresponding to lines 1238–87 of M.)

111	the	your
434	'blesse me:	he
451	I	And
468	I haue don't then:	Be it thus then.
525	*Paw.* with all speed (Sir) — *Exit*	Goe, be gon:
606	I	And
625	not easely	not
689–90	*through the submissiue acknowledgement of your disobedience*	*through your submissiue acknowleg-ment*
863	me:	me well enough

1247	*wh. Q.* No Rescue: no Deliverer?	there is no remedie
1506	*Bl. Q.*	*Bl. K.*

Most of the changes are minor syntactical adjustments, or insertions to fill out a deficient line of verse. At three points (lines 525, 1247, and 1506) a character is deprived of a speech, but as in each case the character concerned appears and speaks elsewhere in the scene, the alteration clearly has no theatrical significance.

In the third group both of the divergent readings are supported by other witnesses, and Crane himself may not necessarily be responsible for the difference. These are not listed here. In the final group, the reading of one manuscript agrees with Trinity and the reading of the other is unique, and is therefore likely to be an error on Crane's part. As this material is helpful in determining Crane's accuracy as a scribe, a list is given below. Readings in the left-hand column agree with Trinity, while those in the right-hand column are unique. (This list excludes the portions of L which are omitted in M, and ignores simple mistranscriptions in M, such as occur at lines 683, 851, and 1060.)

145	not	no L
216	lock	shut L
231	that	which L
268	Virgin	Lady M
271	would	will M
297	on	of L
309	publique	privat M
311	keepe	be M
370	and	or M
440	my	thy L
447	resist me	resist M
448	help: oh help:	help: help: oh help. M
600	Eare	eares M
756	that	her M
770	fowle	Falce M
869	of	in M
1052	work'd	wrought M
1070	I, you	you M
1097	'pray	I'pray M
1111	powre	*Powres* M
1112	Side	State M
1130	*Pound*	*poundes* L

1154	(Sir)	*omitted* L
1211	what a	what L
1286	sound	strong M
1391	the	that M
1418	now you maye goe T	You may now goe L You may goe now M
1557	our[2]	or L
1640	*Fore-heads*	heads L

The most distinctive feature of M is its 'massed entries'. It should perhaps be made clear that in Crane's time there were two ways of dividing plays into individual scenes. In what may be termed the English system, a scene ended when the stage was completely empty of characters. Within a single scene there could be frequent alterations in the number of characters on stage, and it was therefore essential to indicate, by means of entrance and exit stage-directions, when a new character entered, or a character already present left the stage. With the continental or neoclassical system, a new scene began whenever a character entered or left; the stage-direction at the head of each scene was a simple list of the characters taking part in the scene, and it was unnecessary to use words such as 'Enter' or 'Exit'. Crane's practice in M was an awkward compromise that did not strictly follow either of these conventions. He retained the English scene-division found in other texts of *A Game at Chess*, but gathered together all the entrance stage-directions which would normally be scattered throughout the scene and wrote them out as a single long stage-direction at the beginning of the scene.

On the reasonable assumption that Crane derived M from a transcript of his own that was very similar to L, this would not have been difficult to do. The appearance of L suggests that Crane prepared it in three stages. He first ruled out the box-lines on pairs of conjugate leaves. He then wrote out, in one operation, a page of text including speech-prefixes, spoken text, and exits in his normal size of italic. As a final stage, he added in the right margin, usually in a distinctive bold italic, or a mixture of bold and normal italic, the entrances within the scenes and a number of fairly elaborate stage-directions. Entries were normally preceded by a heavy dash to indicate their precise location. Occasionally the space for these insertions was inadequate, and they were crammed into a corner (e.g. pages 29 and 35 of L). Crane apparently became aware of this, and on page 84 of L (transcribed below) he deliberately kept short the first two lines of the Latin oration spoken by the Black Bishop's Pawn, in order to leave room for an important stage-direction. In all cases, however, the marginal stage-directions stood out clearly on the page, and it would have been easy for Crane to turn over the pages of the manuscript he was working from and put together the entrances. One small piece of evidence to suggest that he worked in this way is the heavy dash in M to the right of line 247. Crane very rarely used a dash at the end of a line simply as punctuation, and its presence here might indicate that Crane was copying from a manuscript that, like L, had an entry for the White Bishop's Pawn at this point; he started to write it in, and then realized his mistake.

Crane's method in preparing 'massed entries' was to omit the word '*Enter*', and substitute '*Then*' for entries other than the first in the scene. Compare, for example, lines 88–92 of M with the corresponding stage-directions in L:

Enter ye white-Queenes Pawne, & ye Black Queenes-Pawne.
Enter ye Black B^{ps}. Pawne.
Enter ye white B^{ps}. Pawne.
Enter ye Black Knights Pawne.
Enter Bl. K^{t}.
Enter ye white King's Pawne.

Crane was not, however, strictly consistent, and did use the word '*Enter*' at lines 55 and 1004 of M, as well as the odd form '*Comes*' at line 928. He included exit-directions at ten points within a scene, always preceded by a dash (lines 143, 452, 480, 644, 707, 851, 865, 1000, 1173, and 1237), and also at the end of each scene, where he could in fact have omitted them without undue confusion. In addition, M has '*Noice within*' at line 449 (cf. lines 1415, 1420, and 1427), and more elaborate marginal stage-directions at lines 55–8, 826–8, 926–9, 1317–20, and 1634–7.

The stage-directions in M show various anomalies. There is no massed entry for the Induction; strictly speaking, the separate entrance for the two houses at lines 55–8 should have been added to the opening entrance at line 2. (Contrast this with Crane's more consistent procedure at lines 509 and 664.) Scenes where the same character entered twice were a problem: Crane would have found it very cumbersome to give full details within the massed entry. In II. ii L has a re-entry for the Fat Bishop's Pawn at the equivalent of line 546. This was omitted in M, but as the pawn merely brought in books without speaking the omission is understandable. Crane was obliged, however, to provide a fresh entry for the Black Bishop's Pawn at line 1004, though not in the bold italic he would normally have used. The same character's ceremonial entry in Act III, lines 926–9, as part of the trick played with the 'magical glass', should have been placed after the other entries at line 664, but in L and M this act forms a single long scene because of the omission of the comic scene with three pawns which divides it into two parts in all other texts, and Crane may have felt that to omit the direction at this distance from the beginning of the scene could have confused the reader.

Crane made one error in the massed entry at lines 508–9 by failing to mention the White Queen's Pawn, who appears at line 581. L has an entrance for her at the appropriate point, and also provides exits at the equivalent of lines 283 and 374 which are not found in M. Omissions, however, are not exclusive to M. The exits at lines 452 and 480 of M are not found in L, or in any other manuscript of the play, and M has an entry for the Black Queen's Pawn at line 1341 which L fails to provide later in the scene, presumably through a simple error.

It is hard to explain why Crane chose to present M's stage-directions in the way described above; it has no theatrical purpose, and fails to clarify the action for a reader who is ignorant of the play. The massed entries found here and in certain First Folio comedies were used by some scholars in the past to justify the theory of 'assembled texts', but this theory is now so generally discredited that it seems unnecessary to discuss it in detail.[14] Possibly Crane (or Middleton) considered that an approach towards the continental system, used in the printing of classical drama and in most of Ben Jonson's plays, would enhance the dignity of the work by making it seem more purely literary and less theatrical. The fact that M is heavily abbreviated may also be relevant; it was perhaps easier to make cuts if as many as possible of the marginal stage-directions were moved to the beginning of each scene.

Approximately one-third of the full text of *A Game at Chess* was omitted from M. (Any attempt to give precise figures in terms of line-numbers would be misleading, as the omissions frequently consist of part-lines, phrases, or single words.) The cutting was not uniform: Acts I and V lost about a fifth of their lines, whereas Act II, the longest in the play, lost more than half. As R. C. Bald remarks (ed. cit. 29): 'The cuts have been made with considerable skill, and if there were no other texts one would never suspect that so many lines had been omitted.' Line 429 of M, for example, is in other texts a reply to a quite different remark by the White Queen's Pawn which was cut from M, but it reads as a perfectly appropriate response to lines 426–8. Between lines 743 and 744 of M a passage of seventeen lines was omitted in which the Black Knight boasted of his powers of memory; because of this digression the Black Knight had to repeat his question about which of his plots had been discovered, and lines 744–6 were originally a reply to this second question.

The general policy shown in making the abridgement was to leave out repetitions and digressions, but to omit nothing whose removal would obscure the

[14] See W. W. Greg, *The Shakespeare First Folio* (Oxford, 1955), pp. 156–8, and F. P. Wilson, *Shakespeare and the New Bibliography*, rev. edn. (Oxford, 1970), pp. 74–7.

plot or satirical intentions of the play. (Indeed, occasionally the abridgement might be said to draw attention to structural weaknesses in the full text.) II. i, the attempted seduction of the White Queen's Pawn, was the most heavily cut scene of all, with more than half removed, yet still managed to make its point effectively. Small episodes were skilfully chosen for removal: the brief appearances of the Black Knight's Pawn at the end of II. i and the beginning of IV. i, to complain of his pangs of conscience, could safely be omitted because his state of mind was made clear enough at the beginning of IV. ii. Similar omissions are the account of the activities undertaken by the Jesuits *in voto* (I. i. 53–68; 157–72, between lines 135 and 136 of M) and the recriminations of the black pieces placed in the bag at the conclusion (V. iii. 203–37; 2405–2440, between lines 1641 and 1642 of M). Cutting did, however, sometimes produce lines that are metrically clumsy. Often a cut began and ended in the middle of a verse-line, and the resulting part-lines were spliced together. The result was sometimes very unrhythmical (e.g. lines 415, 438, 566, 629, and 769). In a few places relineation after abridgement led to awkwardness (e.g. lines 602, 709, and 790).

It is not clear who carried out the process of abridgement, Crane himself or Middleton. The care with which it was done would seem to point to Middleton, but Crane was a moderately competent versifier who knew the play well and was evidently trusted by Middleton, so his responsibility is not out of the question.[15] We can only speculate on the motives for preparing an abridged version. It is unlikely that the intention was to produce a simplified form of the play for some kind of clandestine production, since the only characters to disappear completely were the three jesting pawns of III. ii (as in L), and all the stage-business of the full version remained unaltered. The most probable explanation is that Middleton and Crane wanted to make it easier to prepare presentation copies, and the simplest way was to make substantial cuts, but it had to be done as discreetly as possible so that the recipients would not realize how drastically the text had been reduced. The first two lines of Middleton's verse-dedication to Hammond show that at the end of 1624 the play was available only in the form of scarce and desirable manuscript copies, but this obviously ceased to be true as the printed quartos began to appear in 1625.

The transcript of M has been given an independent through-line numbering in the right margin. Cross-references to the Trinity *Game*, edited by Howard-

[15] In 'Shakespeare's Earliest Editor, Ralph Crane', *Shakespeare Survey 44* (Cambridge, 1992), 113–29, T. H. Howard-Hill has argued that the cutting was done by Crane, and that he censored certain passages which he found offensive (see esp. 122–4).

Hill, are inserted in the left margin.[16] The pages selected from L have not been given an independent numbering; cross-references to M have been inserted in the right margin, and to Trinity in the left margin. The dedication page of M, written by Middleton, has not been included among the illustrations because reproductions of it are easily available in the editions of Bald and Howard-Hill. F. P. Wilson has reproduced page 14 of M in the article mentioned in note 8.

[16] In Howard-Hill's edition the marginal line-number 2180 has accidentally been inserted a line too early, thus throwing out all the remaining line-numbers, but to avoid confusion no attempt has been made to correct this error.

BODLEIAN MANUSCRIPT MALONE 25

A GAME AT CHESS (BODLEIAN MS MALONE 25)

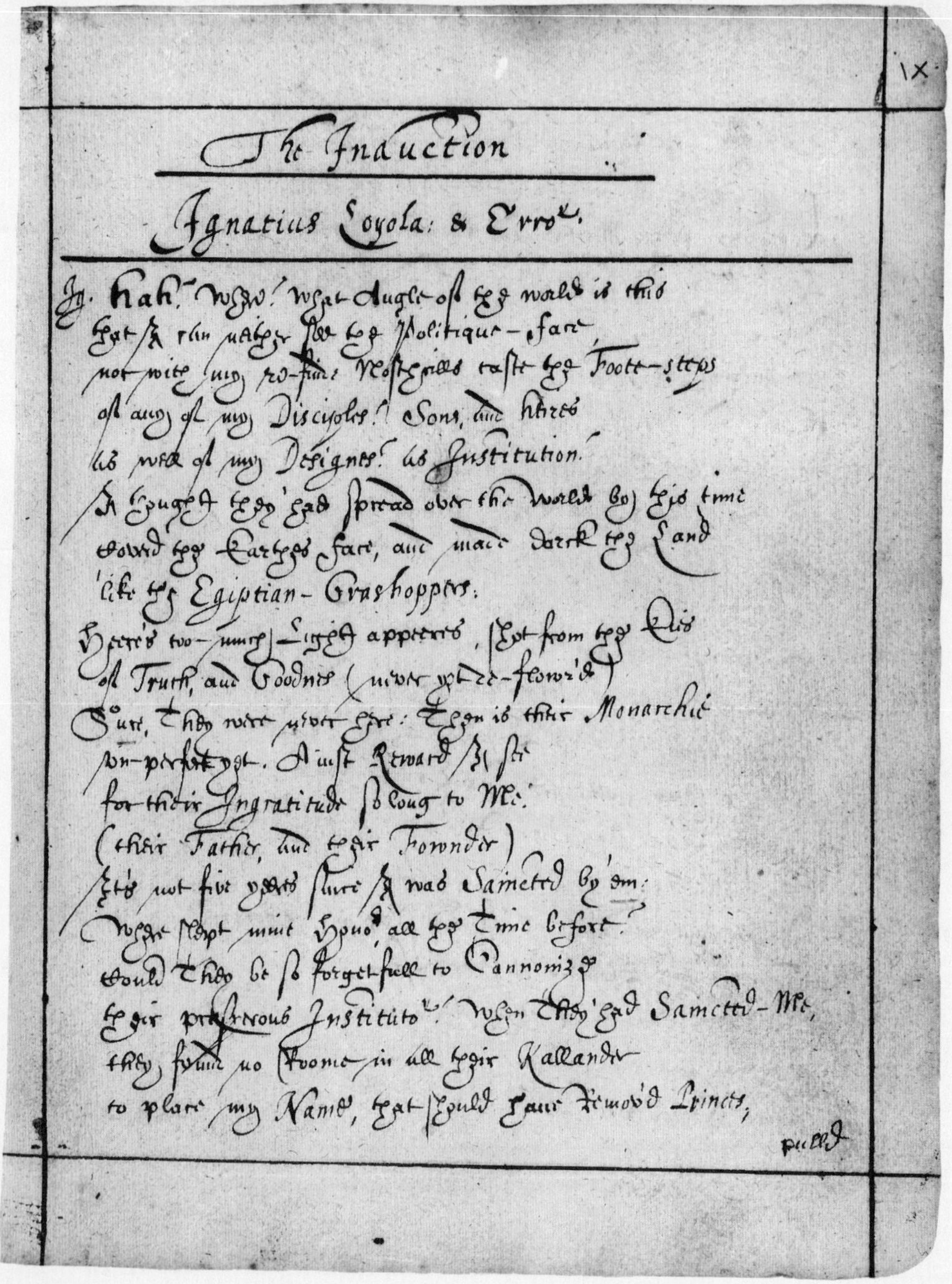

IX

The Induction

Ignatius Loyola: & Error.

Ign. Hah? where? what angle of the world is this
that I can neither see the politique-face,
nor with my re-finde nostrills taste the foote-steps
of any of my Disciples! sonns, and heires
as well of my Designes, as Institution.
I thought they'had spread over the world by this time
coverd the earthes face, and made darke the land
like the Egiptian-Grashoppers.
here's too-much-light appeares, shot from the eies
of Truth, and Goodnes (never yet de-flowr'd)
sure, they were never here: then is their Monarchie
vn-perfect yet. a iust Reward I see
for their Ingratitude so long to Me:
(their Father, and their Founder)
tis not five yeares since I was Sainted by 'em:
where slept mine honor all the time before?
could they be so forget-full to Cannonize
their prosperous Institutor? when they'had Sainted-Me,
they found no roome in all their Kallander
to place my Name, that should have remov'd Princes,
pulld

A Game at Chess (Bodleian ms Malone 25)

Fat B. Then for their Sanctimonious Queenes Surprizall
in that State-puzzell, and distracted hurry,
trust my Arch-Subtletie with it.

Bl. K. oh Eagle-speede,
Never was Game more hopefull of our Side.

Bl. Kt. If Bishop Bull-Beef, be not Snap'd at next Bout,
(as the Men stand) Ile never trust Art more — Exeunt

Scea. Tertia.

A Domb showe. The Bl: Queenes Pawne (with a Taper) Conducts ye wh. Queenes-Pawne (in her night-attire) into one Chamber; Then Conuaies ye Bl. Bps. Pawne, into an other Chamber, So puts out ye Light, & followes him.

Scea. Quarta.

The white-Knight, & wh. Duke; yn ye Bl. Knight: Then ye white-Queene: ye Fat-Bishop: ye white Bishop; & ye wh. King.

wh. Kt. True Noble Duke, faire Vertues most Endear'd-one,
let us prevent their Ranck Insinuation
with Truth of Cause, and Courage; meete their Plotts
with Confident Goodnes, that shall strike 'em grovelling.

Sir

A Game at Chess (Bodleian ms Malone 25)

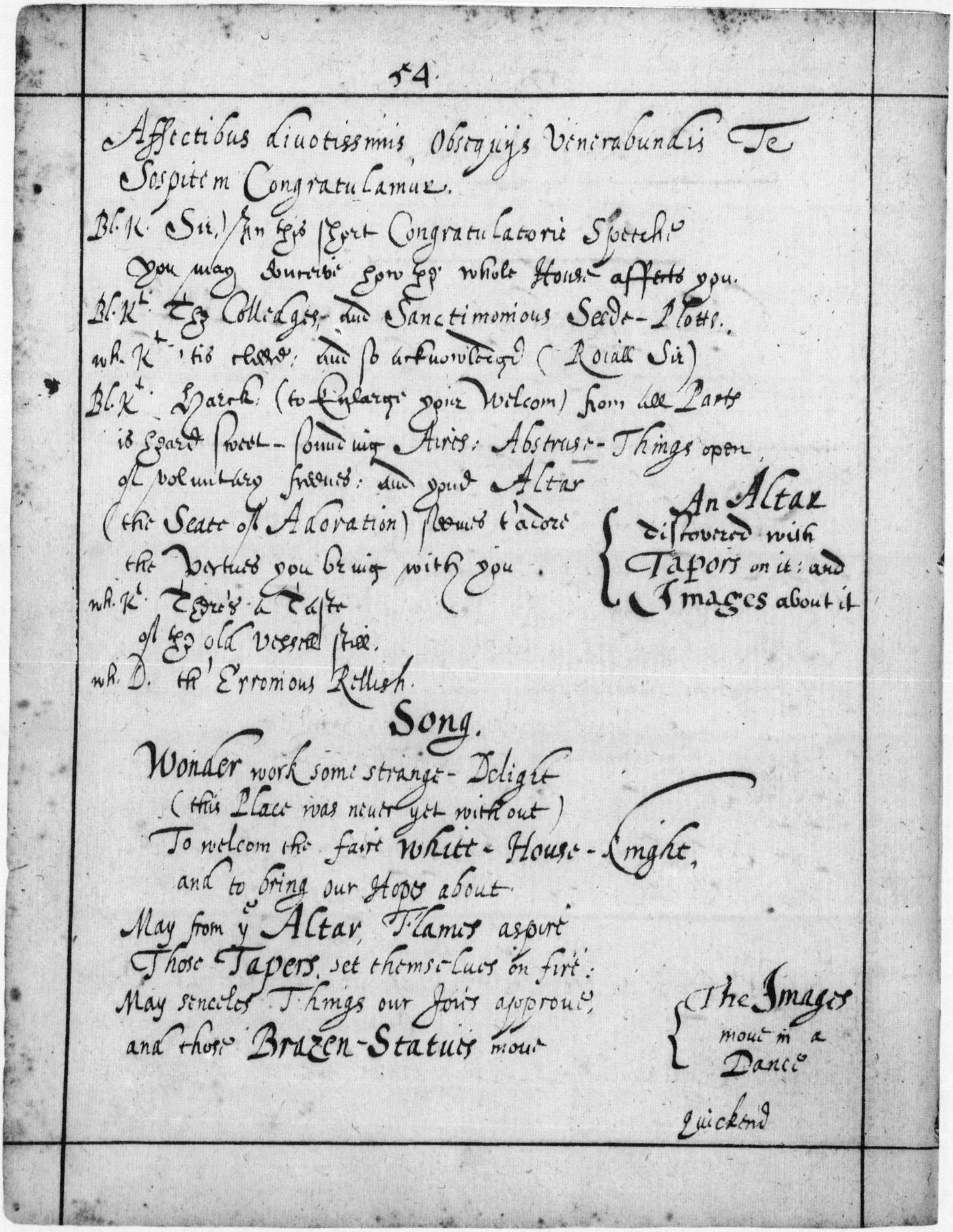

54

Affectibus diuotissimis, Obsequijs Venerabundis Te
Sospitem Congratulamur.

Bl. K. Sir, In this short Congratulatorie Speeche
You may discerne how the whole House affects you.
Bl. Kt. The Colledges, and Sanctimonious Seede-Plotts.
wh. Kt. 'tis cleere; and so acknowledgd (Roiall Sir)
Bl. Kt. Harck: (to Enlarge your Welcom) from all Parts
is heard sweet-sounding Aires: Abstruse-Things open
of voluntarie freeness: and yond Altar
(the Seate of Adoration) seemes t'adore
the Vertues you bring with you.
wh. Kt. There's a Taste
of the old Vessell still.
wh. D. the' Erronious Rellish.

An Altar discouered with Tapers on it: and Images about it

Song.

Wonder work some strange-Delight
(this Place was neuer yet without)
To welcom the faire White-House-Knight,
and to bring our Hopes about
May from ye Altar, Flames aspire
Those Tapers, set themselues on fire:
May senceles Things our Joies approue,
and those Brazen-Statues moue

The Images moue in a Dance

Quickend

A GAME AT CHESS (BODLEIAN MS MALONE 25)

1624

A GAME at Chesse.

By
Tho. Middleton

To the worthilie-Accomplish'd
Mr: William Hammond.

This, which nor Stage, nor Stationers Stall can showe,
(The Common Eye maye wish for, but ner'e knowe)
Comes in it's best loue wth the New-yeare forth,
As a fit Present to the Hand of Worth.

A Seruant to youre
Vertues,

T. M.

The Induction

Ignatius Loyola: & Erro^{r}*.*

Ig. ***hah***? Where? What Angle of the world is this
that I can neither see the Politique-Face,
nor with my re-finde Nosthrills taste the *Foote-steps*
of any of my *Disciples*? *Sons*, and heires
as well of my *Designes*? as *Institution*?
I thought they'had spread over the World by this time
Coverd the Earthes Face, and made darck the *Land*
like the *Egiptian-Grashoppers*:
Heere's too-much Light appeeres, shot from the Eies
of *Truth*, and *Goodnes* (never yet de-flowr'd)
Sure, They were never here: Then is their *Monarchie*
vn-perfect yet. A iust *Reward* I see
for their *Ingratitude* so long to Me.
(their *Father*, and their *Fownder*)
It's not five yeeres since I was *Saincted* by 'em:
Where slept mine Hono^{r}, all the Time before?
Could They be so forgetfull to Cannonize
their prosperous *Instituto*^{r}? When They'had *Saincted*-Me,
they found no Roome in all their *Kallander*
to place my *Name*, that should haue Remov'd *Princes*;

pulld

10 *like*] preceded by fleck in paper resembling an apostrophe

pull'd the most Eminent *Prelats*, by the Rootes vp
for my deere Com̃ing; to make-way for Me.
,Let every petty-Martir, and *Saint-Homilie*,
Roch. Main, and *Petronell*. (*Itch*, and *Ague* Curo^{rs})
Your *Abbesse Aldegund*, and *Cunigung*,
the *widow-Marcell*: *Parson Policarpe*,
Sicelie and *Vrsula*, all take-place of Me.
And but for the *Bis-sextile*, or *Leape-yeere*
(and that's but one in Three) I fall by chaunce
into the *Nine* and *twentith* daie of *Februarie*
there were no Roome els for Me: 'See their Love,
(their Conscience too) to thrust Me (a *Lame Soldier*)
into *Leape yeere*? My wrath's vp: and (methincks)
I`could with the first Sillable of my *Name*
blow-vp their *Colledges*: Vp *Erro*^{r}, wake.
Father of Superarrogation, Rise:
It is *Ignatius* calls Thee (*Loyola*)
Er. What haue you don? oh, I could sleep in *Ignorance*
im̃ortally: the Slomber is so pleasing.
I saw the Bravest-*Setting* for a *Game* now
that ever mine eie fixd on:
Ig. What *Game* 'prethee?

The

25 *,Let*] sic 34 *thrust*] *h* touched up

Er. The Noblest *Game* of all: *a* ***Game at Chesse***
betwixt *Our Side*, and the *white-house*. The *Men sett*
in their iust *Order*, ready to goe to't.
Ig. were any of my *Sons*, placd for the *Game*?
Er. yes: and a *Daughter* too: a *Secular-Daughter*
that Plaies the *Black-Queenes-Pawne*: He the *Black-Bishop's*.
Ig. yf ever *Powre* could show a Maistery in Thee
Let it appeare in this
Er. 'tis but a Dreame,
a Vision, you must thinck.
Ig. I Care not what:
so I behold the *Children* of my *Cunning*
and see what *Ranck* they keepe.
Er. you haue your wish

Enter (seuerally) ***ye white-house, & Black House*** (***in order of the Game***

Behold, there's the full *Nomber of the Game*
Kings, and their *Pawnes*, *Queenes*, *Bishop's*, *Knights* and *Dukes*.
Ig. *Dukes*? They are calld *Rookes* by some
Er. Corruptively:
La Roch, the word, Custode de la-Roch,
The *Keeper* of the *Forts*: In whom both *Kings*
repose much Confidence: and for their Trust-sake,
Courage, and worth, doe well deserve those Titles.

The

47 *ready*] *e* altered from *a*

Ig. The Answeare's high: I see my *Son*, and *Daughter*.
Er. Those are Two *Pawnes*: the *Black-Queenes*, and the *Bishop's*.
Ig. *Pawnes* argue but poore Spirits, and slight *Preferments*.
    not worthie of the Name of my *Disciples*.
    yf I had stood so nigh, I would haue Cutt
    that *Bishop's*-Throat, but I'would haue had his *Place*,
    and told the *Queene* a Love-Tale in her eare,
    would make her best Pullsse daunce: There's no Elixer
    of Braine, or Spirit amongst'em:
Er. why, would you haue them *Play* against themselues?
    that's quight against the *Rule* of *Game* (*Ignatius*)
Ig. pish: I would Rule myself: not obserue Rule.
Er. why, then you'ld play a *Game* all by your self.
Ig. I would doe any thing to *Rule* alone,
    'tis rare, to haue the World raignd-in by One.
Er. See Them anon: and Mark, them in their *Play*.
    Observe (as in a *Daunce*) they glide away.
Ig. Oh, with what Longings will this Breast be tost,
    vntill I see this *Great-Game*, won, and lost? – *Exeunt*/

82 *Mark*,] comma may be accidental pen-mark

98

Actus Primus.

Sce^{a}. prima.

99–101 ***The white-queenes, & y^{e} Black-queenes Pawnes. Then y^{e} Black Bishop's Pawne: Then y^{e} whi: Bishop's Pawne, & y^{e} Bl. Knights Pawne,***
\+ ***Then y^{e} black-knight, Then y^{e} wh. Kings Pawne.***

102–3 *B^{l}. Q^{s}. P.* ***I neuer*** see that Face, but my pittie rises
When I behold so cleere a Master-peece
of Heavens *Art*, wrought out of Dust, and Ashes,
And at next Thought, to give her lost eternally
in being not *Ours*, but the Daughter of *Heresie*.
my *Soule* bleeds at mine eies.
wh. Q^{s}. P. Wher should Truth speake
110 if not in such a Sorrow? Theis are Teares plainely.
112 What is my Peace to her, to take such paines in 't.
118 It's doubtles, a great Charitie, and no Vertue
could wyn me surer?
120 *B^{l}. Q^{s}. P.* Blessed Things prevaile with't
yf ever Goodnes made a gratious promise

it

99 *Wher*] *e* altered from *a*

it is in yonder looke. What litle paines
would Build a *Fort* for Vertue, to all Memorie
in that sweet Creature, were the Ground-work firmer?
wh. Q^s. *P.* It hath byn all my Glory to be firme
in what I haue professd.
127 B^l. Q^s. *P.* that is your Enemie:
130 Your Firmenes that way, makes you more Infirme
for the right Christian Conflict. There I spide
a zealous *Primatiue* Sparcle but now flew
133–4 from your Devoted eie, able to blow vp all the *Heresies*
135 that ever sat in Councell with your Spirit:
138 And here comes He, whose Sanctimonious Breath
will make that Sparck a Flame. List to him (*Virgin*)
140 at whose First Entrance *Princes* will fall prostrate,
Woemen are weaker Vessells.
wh. Q^s. *P.* By my Penitence
a Comely *Presentation*, and the *habit*
to Admiration reverend.
B^l. Q^s. *P.* But the hart (*Ladie*) so meeke
that, as you see good *Charitie* pictured still
with yong-ones in her Armes, so will he Cherish
all his yong Tractable, sweet Obedient *Daughters*
even in his Bosom (in his owne deere Bosom)

I

3.

150 I am myself a *Secular-Iesuite*
(as many *Ladies* are of wealth, and Greatnes)
A Second sort, are *Iesuites* in *Voto*
giving their *Vow*, into the *Father-generall*
(that's the *Black-Bishop* of our *House*, Whose *Pawne*
this *Gentleman* now stands for) to Receive
156 the *Colledge-habit* at his holy pleasure.
173 This *Misterie* is too deep yet for your Entrance,
and I offend to sett your *Zeale* so back,
Check'd by *Obedience*, with desire to hasten
your Progresse to *Perfection*: I Com̄it you
to the Great Workers hands; To whose Grave worth
I fit my reverence, as to you my Wishes.
B^{l}. *B*^{s}. *P*. Doe you find her supple?
180 *B*^{l}. *Q*^{s}. *P*. there's a litle passage – *Exit*
181 *B*^{l}. *B*^{s}. *P*. Let me Contemplate:
185 Amongst the Daughters of Men, I haue not found
a more Catholicall *Aspect*: That Eie
doth promise Single Life, and Meeke *Obedience*:
Vpon those Lipps (the sweet fresh Buds of youth)
the holy Dew of Praire, lies like a Pearle,
190 dropt from the opening Eie-lids of the *Morne*,

vpon

4.

vpon the Bashfull *Rose*. How beuteously
a gentle *Fast*, (not rigorously imposd)
would looke vpon that Cheeke? and how delightfully
the Curteous phisick of a Tender *Penance*
(whose vtmost Crueltie should not exceed
the first feare of a *Bride*) to Beat-downe *Frailetie*
would work to sound health, your long festerd Iudgement?
and make your *Merit*, which (through Erring Ignorance)
appeeres but spotted *Righteousnes* to me
far cleerer then the Inocence of *Infants*?
wh. Q^{s}*. P.* To that good work I bowe: and will becom
Obedience humblest Daughter, since I find
th'assistance of a sacred Strength, to aid Me,
The Labour is as easie to serve Vertue
the right way: since 'tis She I ever seru'd
in my Desire, though I transgresd in *Iudgement*.
B^{l}*. B*^{s}*. P.* That's easely *Absolu'd* amongst the rest.
You shall not find the Vertue you serve now
a sharp, and Cruell *Mistris*: Her Eare's open
to all your Supplications: you may boldly
and saffely let in the most secreat *Sin*
into her knowledge, which (like Vanishd Man)

neuer

153 *delightfully*] *ll* blotted

5.

neuer returnes in to the World agen.
Fate locks not vp more trulier.
wh. Q^{s}*. P.* To the guiltie
that may appeere some Benefit
B^{l}*. B*^{s}*. P.* who's so Inocent
that never stands in need on't in some kind?
Yf every Thought were blabd, that's so confesd
220 the very Aire we breath would be vnblesd.
Now to the Work indeed, which is to Catch
her *Inclination*: (that's the speciall vse
223 we make of all our Practise in all *Kingdomes*)
229 *Daughter* the sooner You dispeirse your Erro^{rs},
230 the sooner you make haste to Your *Recouerie*.
231/235 You must part with'em, – Resolue you thus Far (Lady)
the privatst thought, that runs to hide it self
in the most secreat Corner of your hart now
must be of my Acquaintance so famillierly
Never She-Frend of your Night-Councell neerer.
240 *wh. Q*^{s}*. P.* I stand not much in feare of any Action
guilty of that Black-time; (Most Noble *Holines*)
I must Confes, as in a Sacred Temple
(throngd with an Auditorie) some come rather
to feed on Humaine Obiect, then to taste

of

of *Angells Foode*:
So in the Congregation of Quick Thoughtes
(which are more infinite, then such *Assemblies*)
I cannot (with Truthes saffetie) speake for all:
Some haue byn Wanderers, Some fond, Some sinfull.
But those found ever but poore Enterteinement
they had small Encouragement to Come againe.
The Single-Life, which strongly I profes now.
(heaven pardon me) I was about to part from.
B^{l}. B^{s}. *P*. then You haue passd through *Loue*?
wh. Q^{s}. *P*. But left no Stayne
in all my passage (Sir) no print of wrong
for the most Chaste Maid, that may Trace my Foote-stepps.
B^{l}. B^{s}. *P*. how came you off so cleere?
wh. Q^{s}. *P*. I was dischargd
by an Inhumaine Accident, Which Modestie
forbids me to put any Language to.
B^{l}. B^{s}. *P*. how you forget Yourself? All Actions
clad in their proper Language (though most sordid)
my Eare is bound by Dutie, to Let in,
and lock vp everlastingly. Shall I help You?
He was not found to answeare his Creation.

A

A vestall Virgin, in a slipp of Praire
could not deliuer Mans-losse modestlier.
'twas the *white-Bishop's Pawne*?
270 *wh.* Q^s. *P.* the same (Blessd Sir).
B^l. B^s. *P.* An Heretique well pickelld.
wh. Q^s. *P.* by base Treacherie
and Violence prepard by his Competito^r
(the *Black-Knights Pawne*) whom I shall ever hate for't.
B^l. B^s. *P.* 'twas (of Revenges) the Vnmanliest way
that ever *Riuall* tooke: a Villany
277 that (for your sake) I'll nere *Absolue* him off.
282 It seemes then you refusd him for *Defect*?
Therein you stand not pure from the *Desire*
that other Women haue in ends of *Marriage*.
pardon my Boldnes, if I sift your Goodnes
to the last Graine.
wh. Q^s. *P.* I reverence your paines (Sir)
and must acknowledge, *Custome* to enioy
what other Women Challenge, and posses
290 more ruld me then *Desire*, for my *Desires*
dwell all in *Ignorance*; and I'll neuer wish
292 to know that fond way may redeeme 'em thence.
305 B^l. B^s. *P.* *She's* impregnable:

My

238 *Ignorance*] preceded by pen-rest

8.

310 My old Meanes I must fly to: (yes 'tis it)
'please you pervse this Small *Tract of Obedience*?
'twill help you foreward well.
wh. Q^s. *P.* (Sir) that's a Vertue
I haue ever thought on, with especiall *Reuerence*.
B^l. B^s. *P.* you will Conceive by *That*, my *Powre*, your *Dutie*.
316 *wh.* Q^s. *P.* The knowledge wilbe pretious of both (Sir) –
318 *wh.* B^s. *P.* What makes yond Troubler of all Christian Waters
so neere that Blessed Spring? But that I know
320 her *Goodnes* is the *Rock* from whence it issues
vnmoveable as *Fate*, 'twould more afflict me
322 then all my Suffrings for her.
328 B^l. B^s. *P.* Behold (*Lady*)
the Two Inhumaine Enemies: the *Black-Knights Pawne*,
330 and the *white-Bishop's*: (the *Guelder*, and the *Guelded*.)
331 *wh.* Q^s. *P.* There's my Greif, my Hate.
333 B^l. K^{ts}. *P.* What? in the *Iesuites* Fingers?
I'll give my part now, for a Parrots Feather.
She neuer returnes Vertuous, 'tis impossible
I'll vndertake more Wagers wilbe laid
vpon a Vsurers returne from hell
then vpon hers from Him now: I haue byn guilty
of such base Malice, that my Very Conscience

shakes

340 shakes at the memorie of: And when I looke
to gather *Fruit* find nothing but the *Sauin*-Tree
too frequent in *Nuns* Orchards: (and there planted
by all Coniecture, to destroy fruit rather.)
344 I wilbe resolu'd now: (Most noble Lady)
353 *B^{l}. B^{s}. P.* Son of offence, forbeare; Goe, sett your evill
before your eies: a penetentiall Vesture
will better becom You: some Shirt of haire.
B^{l}. K^{ts}. P. And you a Three pound Smock, stead of an *Alb*,
an Epiceane *Cassible*: This holy ffellon
robbs saffe, and close: I feele a Sting, that's worsse too.
White-Pawne, ha'st so much Charitie, to accept
360 a *Reconcilement*? Make thine owne Conditions
For I begin to be extreemely burthend.
wh. B^{s}. P. No Truth, or Peace of that *Black-house* protested
is to be trusted. But for hope of Quittance,
and warnd by Diffidence, I may Entrap him soonest
I admit Conference.
B^{l}. K^{ts}. P. It is Noblenes
that makes Confusion cleave to all my Meritts.
368 *B^{l}. B^{s}. P.* That *Treatise* will instruct you fully.
370 *B^{l}. K^{t}.* Soe: Soe:
The Busynes of the Vniuersall *Monarchie*
goes forward well now: The Great *Colledge Pot*.

that

274 *saffe*] ^{2}*f* touched up

that should be alwaies boyling with the Fewell
of all *Intelligencies* possible
through the *Christian Kingdomes*. Is this Fellow
our *Prime Incendiarie*? and one of Those
that promisd the *white-Kingdome* seaven yeare since
378 to our *Black-house*? Put a new *Daughter* to him,
the Great Work stands.
379–80 He minds nor *Monarchie*, nor *Hierarchie*
(Diviner *Principalitie*) I haue bragd lesse
381 but I haue don more, then all the *Conclaue* on'em.
384 And what I haue don, I haue don Facetiously
with pleasant Subtletie, and Bewitching Courtship.
Abusd all my Beleevers with Delight
They tooke a Comfort to be Cozond by me.
To many a Soule, I haue let in Mortall poyson
whose Cheekes haue crackd wth Laughter to receive it
390 I could so rowle my Pills in Sugerd Sillables
and strew such kindly Mirth ore all my Mischeifes.
They tooke their Bayne in way of *Recreation*
as Pleasure steales Corruption into youth.
He spies me now: I must vphold his *Reuerence*
395 (especially in privat) though I know
397 *Bacchus*, and *Venus-Chitt* are not more Vitious

Blessings

B^l. B^s. P. Blessings Accumulation, be With you (Sir)
B^l. K^t. Honors Dissumulation be your due (Sir)
wh. Q^s. P. How deep in Dutie his observance plundges?
his Charge must needs be reverend.
B^l. B^s. P. I am *$Confesso^r$*.
to this *Black-Knight* too: You see Devotion's fruitfull.
Sh'ath many Sons, and Daughters.
B^l. K^t. I doe this the more
t'amaze our *Aduersaries* to behold
the *Reuerence* we give theis *Guytinens*:
and to beget a sound opinion
of *Holines* in them, and *Zeale* in vs.
Asalso to Invite the like obedience
in other *Pusills*, by our Meeke example.
Soe: Is your *Triffle* vanishd?
B^l. B^s. P. *Triffle* call You her? 'tis a good *Pawne* (Sir)
Sure she's the Second *Pawne* in the *white-house*
and to the opening of the *Game*, I hold her.
B^l. K^t. I: you hold well for that: I know Your *Play* of old
Yf there were more *Queenes Pawnes*, you'ld ply your *Game*
a great deale harder: (Now Sir, we are in privat)
But what for the *Maine*-work? the Great *Existence*?
the *Hope Monarchall*?

It

312 *Dissumulation*] ^1u possibly has an accidental extra minim 325 *Triffle*] ^2f touched up
327 *house*] followed by pen-rest

B^l. B^s. P. It goes on in this.
B^l. K^t. In this? I cannot see't
B^l. B^s. P. you may deny soe
A Dialls Motion, 'cause you cannot see
the Hand move, or a Wind, that Rends the Cedar.
B^l. K^t. Where stops the Current of *Intelligence*?
Your *Father-generall*, *Bishop* of the *Black-house*
Complaines for Want of Work.
B^l. B^s. P. heer's from all *Parts*
sufficient to employ him: I receiud
a *Packet* from *th' Assistant* fathers lately.
Looke; there's *Anglica*: this *Galica*.
B^l. K^t. I mary Sir)
there's some quick flesh in this.
B^l. B^s. P. Germanica:
B^l. K^t. I thinck they haue seald this with *Butter*.
B^l. B^s. P. This *Italica*.
B^l. K^t. They haue put their *Pens*, the *hebrew* way (Methincks.)
B^l. B^s. P. Hispanica here.
B^l. K^t. Hispanica? Blind work 'tis.
The *Iesuit* hath Writ this with Iuyce of *Lemmans* sure.
It must be held close to the Fire of *Purgatorie*
ere it can be read.

You

B^{l}. *B*^{s}. *P*. you will not loose your *Iest* (*Knight*)
444 though it wound your owne Name.
446 *B*^{l}. *K*^{t}. *Curanda Pecunia.*
B^{l}. *B*^{s}. *P*. Take heed (Sir) we are Entrapd: The *white King's Pawne*?
B^{l}. *K*^{t}. He's made our owne (man) half in Voto yours.
His hart is in the *Black-house*: Leave him to Me.
450 Most-of-all *Frends* Endeer'd: pretiously speciall.
wh. K^{s}. *P*. you see my Out-side: but You know my hart (*Knight*)
great difference in the Colour: There's some *Intelligence*:
and as more ripens, so Your *Knowledge* still
shall prove the *Richer*: There shall nothing happen
(beleeve it) to Extenuat your *Cause*,
or to Oppresse her Frends, but I will strive
to Crosse it with my Councell, Pursse, or Powre.
Keep all Supplies back both in Meanes, and Men
that may raise Strength against you: We must part.
460 I dare not longer of this Theame discusse
the *Eare* of *State*, *is quick and iealious.*
B^{l}. *K*^{t}. *Excellent Estimation*, thou art valued
aboue the *Fleete* of Gold, (that Came short home)
Poore-Iesuite-ridden Soule, how art thou Foold
out of thy Faith? from thy *Allegeance* drawne?
466 *which way soe-ere Thou tak'st, Thou'art a Lost Pawne – Ex*^{t}.

363 *Most-*] hyphen may be cross-bar of *t* 365 *Colour*] *l* touched up 377 *Foold*] *d* touched up

468 # *Actus Secundus.*

Sce^{a}. prima.

469–70 ***The white-Queenes Pawne*** (reading) ***The Black***
\+ ***B^{s}. Pawne; Then y^{e} Black-Queenes-Pawne.***
\+ ***Then y^{e} Black-Bishop, & Black-Knight.***

471 *wh. Q^{s}. P.* **And** here agen: *It is the Daughters Dutie*
to Obay her Confessors Commaund in all Things
without Exception, or Expostulation.
It's the most generall *Rule*, that ere I read of.
Yet, when I thinck how boundles Vertue is
Goodnes, and *Grace*, 'tis gently reconcilde
And then it appeeres well to haue the Powre
478 of the *Dispenso^{r}* as vncircumscrib'd.
502 *B^{l}. B^{s}. P.* Sh'ath past the generall *Rule*, the Large *Extent*
of our *Prescription for Obedience*,
and yet, with what Alacritie of Soule
her Eie moves on the Letters
wh. Q^{s}. P. holy Sir
too long I 'haue missd you: oh, your Absence starves me;
508 hasten for Times redemption (worthie Sir)

for

385 *here*] *h* blotted

15.

521 for Vertue sake (good Sir) *Commaund* me something
Make triall of my Dutie in some small Service
and as you find the Faith of my *Obedience* there
then trust it with a greater.
B^{l}. B^{s}. *P.* you speake sweetely:
I doe *Command* you first then.
wh. Q^{s}. *P.* with what Ioy
I doe prepare my Dutie?
B^{l}. B^{s}. *P.* To Meete me,
530 and seale a kisse of Loue vpon my Lip
wh. Q^{s}. *P.* hah?
B^{l}. B^{s}. *P.* At First Disobedient? In so litle too?
how shall I trust you with a greater then?
which was your owne request?
wh. Q^{s}. *P.* 'pray send not back.
36/543 Mine Inocence to wound me: (Sir) If my *obedience*
and your *Commaund*, can find no better way
545 *fond Men Commaund, and Wantons best obay.*
555 Yf this be Vertues path, 'tis a most strange one,
I neuer came this way before.
557 B^{l}. B^{s}. *P.* that's your Ignorance:
562 Your ffeare is wondrous Faultie: Cast it from You:
'twill gather els in time a Disobedience

too

413 *request?*] followed by pen-rest   417 *obay*] *a* blotted

16.

too stubborne for my *Pardon*:
wh. Q^s. *P*. haue I lockd myself
at vnawares, into Sins Servitude
567/571 with more desire of Goodnes? when a Virgin's ruynd
I see the Great work of *obedience*
573 is better then half Finishd.
585 B^l. B^s. *P*. Was that Scorne?
I would not haue it prove soe, for the hopes
of the *Grand Monarchie*: yf it were like it
Let it not dare to stir abroad agen
a stronger Ill, will Coape with't
590 *wh*. Q^s. *P*. he threatens Me.
598 B^l. B^s. *P*. A plaine, and most insufferable Contempt
My Glory, I haue lost vpon this Woman
600 in Freely offring that she should haue kneeld
601/608 a yeere in vaine for: Lay me downe *Reputation*,
before thou stirst: Thy Nice Virginitie
610 is recompence too litle for my Loue,
611 'tis well if I accept of that for both.
622 We must not trust the Policie of *Europe*
vpon a Womans Tongue.
wh. Q^s. *P*. Then take my life,
and leave mine $Hono^r$, for my Guid to heaven.

Take

434 *threatens*] *s* blotted

B^{l}. B^{s}. *P*. Take heed I take not both: Which I haue Vowd
if Longer thou resist.
628 *wh*. Q^{s}. *P*. help: help: oh help.
633–4 B^{l}. B^{s}. *P*. Must Force Confound noice. {*Noice within*
636–7 *wh*. Q^{s}. *P*. I'll venture my Escape vpon all Dangers now.
642 And will discover Thee (*Arch-Hipocrite*)
643 to all the Kindreds of the earth. – *Exit*
646 B^{l}. Q^{s}. *P*. Are you mad?
647 Can Lust infatuat a Man so hopefull?
649 Time. and faire Temper would haue wrought her pleasant.
650 I spide a *Pawne* o'th' *white-house* walk neere vs,
and made that Noice, on purpose to give Warning
652 (for mine owne Turne: Which end in all I work for.)
660 B^{l}. B^{p}. oh, you'haue made noble Work, for the *white house* yonder
This Act will fill the *Aduersaries* mouth,
662 and blow the *Lutherans* Cheekes, till they crack agen.
667 B^{l}. B^{s}. *P*. I dwell not (Sir) alone in this Default
668 the *Black-house* yeilds me *Partners*,
676 B^{l}. *B*. She hath no Witnes then. ?
678 B^{l}. K^{t}. grosse: Witnes?
When went a Man of his Societie
680 to Mischeif with a Witnes?
B^{l}. *B*. Be it thus then.

Away.

462 *this*] *s* altered   466 *his*] *s* altered

18.

682 Away: Vpon the Wings of Speed: Take *Post-horse*;
\+ Cast Thirtie Leagues of earth behind thee sodainely.
683 Leave Letters Ante-dated, with our *house*,
684 Ten daies at least from this.
688 *B^{l}. B^{s}. P*. But (good Sir)
how for my getting forth Vn-spide?
690 *B^{l}. K^{t}*. There's Check agen.
691 *B^{l}. Q^{s}. P*. No: I'll help that.
693 There lies a secreat Vault.
695 *B^{l}. B^{s}. P*. Run for my Cabinet of *Intelligencies*
for feare they search the House: good *Bishop* burne'em rather.
697 I cannot stand to pick'em now – *Exit*.
700 *B^{l}. K^{t}*. Let me see (*Queenes Pawne*.)
701 How formally 'hath pack-vp his *Intelligencies*?
705 Oh, this is the *English-house*: what Newes there troa?
704/706 *Anglica*: Most of theis are Bawdy *Epistles*.
708 Heere's from his *Daughter Blanch, and Daughter Bridget*
from their Saffe Sanctuary in the *white-Friers*.
710 Theis from Two Tender *Sisters* of *Compassion*,
in the Bowells of *Bloomesburie*.
Theis from the *Nunnerie* in *Drurie Lane*
(a Fire: a Fire: good *Iusuitesse*: a Fire)

What

471 *Ante*] *A* altered from *D* 490 *Iusuitesse*] dot of *i* misplaced, giving apparent reading of *Iusiutesse*

What haue You there?
B^l. *B*. A *Note* (Sir) of State *Policie*,
and one exceeding saffe one
B^l. K^t. 'pray let's see it.
To sell away all the Powder in a Kingdome
to preuent Blowing up. (that's saffe, I'll hable it)
720 Heere's a Facetious *obseruation* now.
and suites my humo^r better: He writes here
Some Wiues in *England* will Comitt *Adulterie*
and then send to *Rome* for a *Bull* for their *Husbands*.
B^l. *B*. haue they those shifts?
B^l. K^t. oh, there's no *Femall* breathing
sweeter, and subtler: Here (Wench) take theis Papers
727 scortch'em me soundly.
738 B^l. Q^s. *P*. Feare not in all:
739 I loue *Roguerie* too Well, to let it Fall. – *Exeunt*.|

+ ## Sce^a. secunda.

759–60 ***The Fat-Bishop: & his Pawne: Then ye B^l. B^p. & B^l.***
+ ***Knight. Then ye wh. & Bl. Houses*** (seuerally)

761 *fat B. Pawne*.
Paw. I attend at your *great Holines* Service.

For

495 *all*] ^{2}*l* touched up

fat B. For *great*. I grant yo^{u}: But for greatly *holie*
there the Soile alters: Fat Cathedrall Bodies
haue verie often but leane litle Soules,
Much like the *Ladie* in the *Lobsters* head
a great-deale of Shell, and Garbish of all Colours,
But the pure part, that should take wings, and Mount
is at last gaspe; As if a Man should gape
770 and from this huge Bulk, let Forth a Butter-fly.
773 Are my Bookes printed (*Pawne*) my last *Inuectiues*
against the *Black-house*?
B^{l}. *Paw*. ready for *Publication*
for I saw perfect *Bookes* this Morning (Sir)
fat B. fetch me a Fewe: which I will instantly
778 distribute 'mongst the *white-house*: Goe, be gon:
781 It's a most Lordly life, to raile at ease,
Sit, eate, and ffeed, vpon the Fat of one *Kingdome*,
and raile vpon an other with the Iuyce on't.
I haue writt this *Booke* out of the strength, and Marrow
785 of six and thirtie Dishes at a Meale.
795 But I haue no *Preferment* yet, that's sutable
to the Greatnes of my *Person*, and my *Parts*.
I grant I live at ease: for I am made
the *Master* of the *Beds* (the long *Acre of Beds*)

but

but there's no *Marigolds*, that shutts, and opens,
800 *Flowre-gentles, Venus-Bath, Apples of Loue,*
802 There was a time, I had more such *Drabs*, then *Beds*,
803 now I haue more *Beds*, then *Drabs*.
808 Yonder *Black-knight*, (the *Fistula of Europe*)
Whose Disease once I vndertooke to Cure
810 with a high-*holborne*-halter: When he last
vouchsaffd to peep into my priviledgd *Lodgings*
he saw good store of *Plate* there, and *Rich-hangings*
He knew I brought none to the *white-house* wth me
I haue not lost the vse of my *Profession*
815 since I turn'd *white-house-Bishop*.
818 *B*^{l}. *K*^{t}. Looke, more *Bookes* yet:
Yond Greazie-Turne-Coat, Gurmondizing *Prelat*
820 doth work our *House* more mischeif by his Scripts
(his fat, and Fullsom Volumes)
then the whole Body of the *Aduerse Partie*
B^{l}. *B*. oh, 'twer a *Master*-peece of Serpent Subtletie
824 to Fetch him on this Side agen.
829 *B*^{l}. *K*^{t}. I'll Confound him;
830 on both Sides, for the phisick he prescribd
and the base Surgeon he provided for Me,
I'll tell you what a most Vncatholique Iest

he

22.

---

he put vpon Me once, When my Paine torturd me.
He told me he had found a present Cure for me
wch I grew prowd-on, and obseru'd him seriously.
What thinck you 'twas? Being Execution day
he showd the *Hang-man* to Me, out at *Window*
(the *Common-hang-man*)
B^{l}. B. Insufferable:
840 *B^{l}. K^{t}*. I'll make him the *Baloon* Ball of the *Churches,*
841/844 and both the *Sides* shall tosse him: *Our Second Bishop* absent
845 Which hath yet no Employment in the *Game*,
848 he shalbe flatterd with *Sede Vacante*,
Make him beleeue he Comes into his Place,
850 and that will fetch him, with a Vengance to vs.
854 *B^{l}. B*. no more now (Sir)
856 *wh. King*. This hath byn lookd for long.
fat B. The stronger Sting it shootes into the Blood
of the *Black-Aduersarie*: I am ashamd now
I was theires ever: what a Lump was I
860 When I was lead in Ignorance, and Blindnes.
I must confes I haue all my life time plaid
the ffoole till now.
863 { *B^{l}. K^{t}*. and now he plaies Two Parts
the *Foole*, and *Knaue*.
869 *B^{l}. B*. here comes more Anger.
872 *wh. Q*. Is this my *Pawne*? She that should Guard our *Person*?

or

23.

or some pale figure of Deiection
874 her Shape Vsurping?
876 *wh. Q*^{s}*. P. King* of *Integritie.*
Queene of the same: And all the *House-Professors*
of noble *Cando*^{r}, vncorrupted *Iustice*,
and Truth of hart, through my alone Discovery
880 My life, and Hono^{r} wondrously preseru'd,
I bring into your knowledge with my Suffrings,
fearefull Affrightments, and hart-killing Terro^{rs},
the Great *Incendiarie* of *Christendome*,
the Absolutst Abuser of true Sanctitie
faire Peace, and holy *order*, can be found
in any part o'th' Vniversall Globe,
Who, making meeke Devotion keep the Dore,
his Lipps being full of holy zeale, at First,
889 would haue Com̄itted a fowle *Rape* vpon Me.
891 *wh. K.* A *Rape*? that's fowle indeed; the very sound
to our eares fowler, then the offence it self
93/915 to some *Kings* of the *Earth*; It greives me that
915–16 my knowledge must be taynted with his Infested Name.
oh. rather with thy finger point him out –
wh. Q^{s}*. P.* The Place which he should fill, is void (my lord)
919 his Guilt hath Ceizd him: The *Black-Bishop's Pawne*;

I

941/959 *B*^{l}. *K*^{t}. I Combat with this Cause: And play thus then.
960 Now in the hearing of this high *Assembly*
961 bring forth the Time of this Attempts Conception
970 *wh. Q*^{s}. *P*. yesterdaies haples Evening.
983 *B*^{l}. *K*^{t}. Is it He?
and that the time? Stand firme now to your Scandall,
'pray doe not shift your Slaunder.
wh. Q^{s}. *P*. shift your Treacheries
they'haue worne one Suit too long.
B^{l}. *K*^{t}. That holy Man,
so wrongfully accusd by this *Lost Pawne*
990 hath not byn seene theis Ten daies in theis *Parts*,
992 nay at this instant Thirtie Leagues from hence.
994 *wh. K*. Can you make this appeere?
B^{l}. *K*^{t}. Light is not Cleerer.
By his owne *Letters*, (most impartiall Monarch.)
wh. K^{s}. *P*. how wrongfully may sacred Vertue suffer (Sir?)
998 *B*^{l}. *K*^{t}. (*Bishop*: We haue a Treasure of that Falce hart:
1004 There's an Infallible Staff, and a *Red-hat*
1005/1007–8 reseru'd for you (A Staff that will not breake,
1009 and such a one had your Corruption need of.
1010 there's a State-*Fig* for you now.)
wh. K. Behold all

how

1017 how they Coheare in one: Where setles the offence
Let the ffaults punishment be deriu'd from thence,
We leave her to your Censure.
1020 *B*^{l}. *K*^{t}. most iust maiestie.
wh. Q^{s}. *P*. Calamitie of Vertue: My *Queene* leave me too?
Am I cast off, as th'Olliff casts her *Flowre*?
poore Frendles *Inocence*, art thou left a Prey
to the Devowrer?
wh. K^{t}. No, Thou art not lost.
(Let 'em put on their Bloodiest *Resolutions*)
yf the faire Policie I ayme at prospers.
Thy Councell (Noble *Duke*)
wh. D. For that work cheerefully.
1030 *wh. K*^{t}. A Man for Speed now?
wh. B^{s}. *P*. Let it be mine Hono^{r} (Sir)
1032–3 make me that fflight, that owes her my lifes Service – *Ex*^{t}.
1036 *B*^{l}. *K*^{t}. Let's vse her, as (vpon the like *Discouerie*)
a *Maid* was vsd in *Venice*: Every one
1038 be ready with a *Penance*: Begin *Maiestie*
1045 *B*^{l}. *K*. First I enioyne Thee to a Three daies *Fast* For't.
B^{l}. *Q*. you are too penurious (Sir) I'll make it Fowre.
B^{l}. *B*. I to a *Twelue-howres Kneeling* at one time.
B^{l}. *K*^{t}. And in a Roome filld all with *Aretines-Pictures*.

more

26.

More then the Twice-Twelue *Labours* of *Luxurie*.
1050 Thou shalt not see so much as the Chast *Pomell*
of *Lucrece* Dagger peeping. Nay, I'll punish Thee
for a *Discouerer*: I'll torment thy *Modestie*.
B^{l}. *D*. After that Fowre Daies *Fast*, to th'*Inquisition-house*.
strengthend with Bread, and water, for worsse *Pennance*.
1055–6 *B*^{l}. *K*^{t}. well said (*Duke of our House*) nobely aggravated.
1057 *wh*. *Q*^{s}. *P*. Vertue (to show her Influence more strong)
1058–9 *Fitts me with Patience, mightier then my wrong.* – *Exeunt*,

1060
Actus Tertius.

+
Sce^{a}*. prima.*

1061 ***The Fat-Bishop: Then the B***^{l}***. K***^{t}***. Then his Pawne.***
\+ ***Then y***^{e} ***white, & Black-Houses*** (seuerally.)

1062 *fat B*. ***I know*** my *Pen* drawes Blood of the *Black-house*,
1063 there's never a *Booke* I wryte, but their *Cause* bleedes.
1067 But where is my *Aduauncement* all this while?
1073 To be made *Master* of an *Hospitall*
is but a kind of Diseasd Bed-rid *honor*.
Or *Deane* of the *Poore-Almes-knights*, that Weare *Badges*:

There's

Ther's but Two lazie Beggerly *Preferments*
in the *white-Kingdome*: and I haue got'em both
My *Merit* doth begin to be Crop-sick
1079 for want of other Titles.
1081 *B*^{l}. *K*^{t}. oh, here walkes
his fullsom *Holines*: Now for the Master-Trick
t'vndoe-him everlastingly: that's put home,
and make him hang in Hell most seriously
that Iested with a Halter vpon Me.
fat B. the *Black-knight*? I must looke to my *Play* then:
B^{l}. *K*^{t}. I bring faire Greetings to yor *Reuerend* Vertues,
1088 From *Cardinall Paulus*, your most *Princely Kines-man.*
5/1098 *fat B. reades*:} *Righ reuerend & holy* (Meaning Me:) *your Vnkind*
1098–9 *disobedience to the Mother-Cause, proues at this time the*
-1100 *onely cause of your ill fortune. My present Remoue by*
1101 *generall Election to the Papall Dignitie, had now (auspitiously)*
1102 *setled you, in my Sede Vacante*: (how? had it soe?)
1107 *B*^{l}. *K*^{t}. The *Pill* workes with him.
108–9 *reades*.}. *It is not yet too late (through your submissiue acknowleg=*
0– { *ment) to be Louinglie receiued into the Brotherlie Bosom of*
1 { *the Conclaue.*
1112 This was the *Chaire of Ease*, I ever aymd at.
I'll make a Bon-fire of my *Bookes* im̃ediatly.
All that are left (against that Side) I'll sacrifize

Pack

683 *Righ*] sic

28.

1115 Pack vp my *Plate*, and *Goods*, and steale away.
1121 (*Black-knight*) Expect a wonder ere't be long.
Thou shalt see me one of the *Black-house* shortly.
B^{l}. *K*^{t}. Your *Holines* is merry with the *Messenger*.
1124 'Too happie to be true: you speak what should be
1132 I tell you (Sir) your *Reuerend Reuolt*
did give the fearefullst Blow to *Adoration*
our *Cause* ere felt: It shooke the very *Statues*,
1135 the *Vrnes*, and *Ashes* of the *Sainted sleepers*.
1138 *fat B*. suffizes I am *Yours*; when *They* least dreame on't
Ambition's Fodder, (*Powre* and *Riches*) drawes me:
1140 When I smell *Hono*^{r}, that's the Lock of hay,
1141–2 that leades Me through the world's *Feild* every way. – *Exit*
1143 *B*^{l}. *K*^{t}. Heere's a sweet *Paunch*, to propagate Beleif on:
1145–6 I may Nomber him now 'mongst my Inferio^{r} *Policies*;
1147 But Let me a litle sollace my *Designes*,
with the remembrance of some Brave-Ones past,
1149 to Cherish the Futuritie of *Proiect*.
1156 Who made the Goales fly open (without Miracle?)
and let the *Locusts* out, those Dangerous *Flies*
whose propertie, is to burne Corne with Touching:
The Heritique *Grannaries* feele it to this Mynut;
1160 And now, they'haue got amongst the Cuntry Crops
they stick so fast to the Converted Eares
the lowdest Tempest that Authoritie rowzes

will

29.

will hardly shake 'em off. They haue their Dens
in *Ladies* Cowches; There's saffe Groves, and ffens,
Nay, were they followed, and found out by th'e Scent,
Palme oile will make a Purseuant relent.
Whose Policie was't, to put a *Silenc'd-Muzzell*
on all the *Barking-Tongue-men* of the Time?
Made *Pictures*, that were dombe enough before,
1170 poore Sufferers in that politique *Restraint*?
My light Spleene skipps, and shakes my Ribbs to thinck on't.
Whilst our drifts walke vncensurd, but in thought,
a Whisper, or a Whistle would be questiond,
In the most fortunate Angle of the World
the Court hath held the Cittie by the hornes,
whilst I haue milkd-her: I haue got good Soapes too
from Cuntrie-*Ladies* for their *Liberties*.
From some for their most vainely hopd *Preferments*,
High *Offices* in th'Aire: I should not live
1180 but for this *Mel-Aerium*, this *Mirth-Manna*.
1181 My *Pawne*? how now? the Newes?
1183 *Paw*. expect none very pleasing
1184 that Comes (Sir) of my bringing: I am for sad Things.
1194 (Sir) your *Plot's* discovered.
B^l. K^t. which of the Twentie thousand and Nine hundred
1196 Three-score, and five? 'canst tell?
1218 *Paw*. your last *Brat* (Sir)
begot betwixt the *Black-Bishop*, and yourself,

your

30.

1220 Your *Antedated Letters* 'bout the *Iesuite*,
1222–3 The *white-knights* Policie hath out-stripd yours (it seemes)
1224 Ioynd with th'assistant Councell of his *Duke*,
The *Bishops white-Pawne* vnder tooke the Iorney.
Who (as they say) Dischargd it like a fflight,
1227 I made him for the Busynes, Fit, and Light.
1233 *wh. King*. Set free that Vertuous *Pawne* from all her Wrongs:
Let her be brought with honor to the face
1235/1237 of her Malitious *Aduersarie*: *Noble Chaste Knight*.
1240 This Faire *deliuering Act* Vertue will *Register*
1241 in her *white-Booke* of the *Defence of Virgins*.
1244 And we Embrace (as *Partner* of that *Honor*)
this worthie *Duke* (the Councell of the *Act*,)
1246 Whom we shall ever place in Our *Respect*.
1254–6 Appeere thou *Beutie* of *Truth*, and *Inocence*,
Best *Ornament* of *Patience*; Thou that mak'st
thy *Suffrings* glorious.
1257 *B^{l}. K^{t}*. What makes *She* here?
how dares yond *Pawne* (vn-*Penanc'd*) with a Cheeke
fresh as her *Falsehood* yet, Where Castigation
1260 hath left no pale print of her Visiting Anguish
1261 appeere in this *Assemblie*?
1276 *wh. King*. This *Black-knight*,
1277/1280 Will never take an Answeare: Shew him the Testimony
(Confirm'd by Good-men) how that Falce-*Attempto^{r}*

got

752 *King*] could be read as *Knig* (see Introduction, p. 7)

got but this Morning to the Place from whence
he dated his forgd lines, for Ten daies past?
B^{l}. *K*^{t}. Why may not that Corruption sleepe in this
by some Connyvence? As you haue Wakd in ours
by too rash Confidence?
wh. D. I'll vndertake
this *Knight* shall teach the Devill how to lie
wh. K^{t}. yf Sin were half so wise, as Impudent,
1290 She would nere seeke farther for an Advocate.
1292–3 *B*^{l}. *Q*^{s}. *P*. Now, to Act Treacherie with an Angells Tongue.
Since all's Come out, I'll bring him strangely in agen:
Where is this Iniurd-*Chastetie*? this *Goodnes*?
1296 whose Worth, no Transitorie Peece can Value?
1299 *B*^{l}. *Q*. what? is my *Pawne* distracted?
1300 *B*^{l}. *K*^{t}. I thinck rather
there is some notable *Master-Prize of Roguerie*
1302 this Drom strikes vp for.
1314 *B*^{l}. *Q*^{s}. *P*. I saw this glorious, and most valiant Vertue
1315 fight the most noblest Combat with the Devill.
1318–19 *wh. King*. then thou heardst the Violence intended?
1320 *B*^{l}. *Q*^{s}. *P*. 'tis a Truth
I ioy to Iustefie. I was an Agent
on Vertues-part, and raisd that confusd Noice
1323 that startled him, and gave her Libertie.
1329 *B*^{l}. *K*^{t}. Degenerate.
1330 *B*^{l}. *Q*. Base

Perfidious

B^l. *B*. Perfidious:
B^l. *D*. Traiterous *Pawne*.
B^l. Q^s. *P*. What? are ye all Be-side yourselues?
B^l. K^t. But I:
remember that (*Pawne*)
B^l. Q^s. *P*. May a fearefull Barrennes
blast both my hopes, and Pleasures, if I brought not
her Ruine, in my Pittie: a new Trap
for her more sure Confusion.
1340 B^l. K^t. haue I won now?
1341 Did not I say 'twas *Craft*, and Machination?
1344 B^l. *K*. That *Queene* would I fayne Finger.
B^l. K^t. You are too hot (Sir)
Yf She were Tooke, the *Game* would be *ours* quickly.
My Ayme's at that *white-knight*: Entrap him first
the *Duke* will follow too.
B^l. *B*. I would that *Bishop*
1350 were in my Diocesse: I would soone change his *white-nes*.
B^l. K^t. Sir: I could whip you vp a *Pawne* im̃ediatly
I know where my *Game* stands.
B^l. *K*. do't sodainely:
Advantage least must not be lost in this *Play*.
B^l. K^t. (*Pawne*, Thou art *Ours*.
wh. K^t. He's Taken by *Default*.
by Willfull-Negligence: Guard the *Sacred-Persons*
Looke well to the *white-Bishop*: for that *Pawne*
gaue Guard to the *Queene*, and him, in the Third-Place.

See

33·

1360 *B^{l}*. *K^{t}*. See what sure Peece, you lock your Confidence in,
1363 This *white-nes* vpon Him, is but the Leaprouzie
of pure *Dissimulation*; View him now.
His *Hart*, and his *Intents*, are of *Our Colour*.
wh. *K^{t}*. Most dangerous *Hipocrite*
{ *he appeeres Black underneath*
1367 *wh*. *D*. One made against vs.
1384 *wh*. *K*. You haue him: We can spare him: and his Shame
1385 will make the *Rest*, looke better to their *Game*.
1389 *fat B*. Is there so much Amazement spent on him
1390 that's but half *Black*? (there might be hope of that Man)
But how will this *House* wonder, yf I stand forth
and show a whole one? instantly Discover
one, that's all *Black*? where there's no hope at all?
wh. *K*. I'll say thy Hart, then iustefies thy *Bookes*.
I long for that Discovery.
fat B. Looke no farther then:
Beare witnes all the *House*, I am the *Man*,
and Turne myself into the *Black-house* freely.
I am of this Side now.
1400 *wh*. *K^{t}*. Monster nere matchd him.
1403 *fat B*. Next newes you heare, expect my *Bookes* against you,
Printed at Doway, Bruxells, or Spoletta.
wh. *K*. See his *Goods* Ceizd on:
fat B. (alas) They were all Convaid
last night by water to a Tailors house,
a Frend of the *Black-Cause*.

a

836 *there's*] *h* touched up

wh. K^{t}. a prepard *Hipocrite.*
1410 *wh. D.* permeditated Turne-Coat. – *Ex*^{t}.
fat B. Yes: raile on:
I'll reach you in my *Writings*, when I am gon.
B^{l}. *K*^{t}. Flatter him a while with *Honor*, till we put him
1414 vpon some dangerous Service, and then Burne him.
1417 *fat B.* Now I'll discover all the *white-house* to you.
B^{l}. *D.* Indeed? that will both Reconcile, and raise you.
wh. K^{s}. *P.* I rest vpon you (*Knight*) for my *Aduancement* now.
1420 *B*^{l}. *K*^{t}. Oh: for the *Staff*? (the strong *Staff* that will hold)
and the *Red-hat*, (Fitt for the Guilty Mazard)
Into the *Emptie-Bag*: know thy first way.
1423 *Pawnes that are Lost, are euer out of Play.*
1425 No Replications: you know me well enough:
No doubt ere long, you'll haue more Companie:
1427–8 The *Bag* is big enough: 'twill hold vs all. – *Ex*^{t}.
1429 *wh. Q*^{s}. *P.* I sue to Thee: 'prethee be one of vs:
1430 Let my Love wyn-thee: Thou hast don *Truth* this day,
and yesterday mine *Hono*^{r}, noble Service.
The best *Pawne* in our *House* could not transcend it.
B^{l}. *Q*^{s}. *P.* My pittie flam'd with zeale, especially
when I fore-saw your *Marriage*: then It mounted.
1435 *wh. Q*^{s}. *P.* how? *Marriage*?
1440 *B*^{l}. *Q*^{s}. *P.* yes: yes: you doe *Marrie.*
1441/1443 I saw the *Man*; an absolute handsom Gentleman:

you'll.

851 *permeditated*] sic

35·

1444 you'll say soe, when You see him: Heire to Three *Red-hatts*,
1445 besides his generall Hopes in the *Black-house*.
1450 *wh*. Q^s. *P*. Why, how came you to see, or know this Misterie?
B^l. Q^s. *P*. A *Magicall Glas*, I bought of an *Egiptian*,
Whose *Stone* retaines that Speculative-Vertue
presented the *Man* to Me: Your *Name* brings him:
as often as I vse it: and (methincks)
I neuer haue enough, (*Person*, and *Postures*
are all so pleasing.)
wh. Q^s. *P*. This is wondrous strange:
The *Faculties* of Soule, are still the same,
I can feele no one Motion tend that way.
1460 B^l. Q^s. *P*. We doe not alwaies feele our *Faith* we live by,
Nor ever see our Growth: yet both work vpwards.
wh. Q^s. *P*. 'twas well applide: But may I see him too?
B^l. Q^s. *P*. surely you may: without all doubt, or Feare,
Observing the right vse, as I was taught it:
Not looking-back, or Questioning the Specto^r.
1466 *wh*. Q^s. *P*. that's no hard *obseruation*: Trust it with me:
1522 B^l. Q^s. *P*. This is the *Roome* he did appeere to Me in:
1523 and (looke you) This' the *Magicall Glas*, that showd him:
1538 *wh*. Q^s. *P*. And will the Vsing of my *Name*, produce him?
B^l. Q^s. *P*. Nay, of yours onely: els the Wonder halted;
1540 To cleere you of that Doubt: I'll put the Difference,

in

36.

in practise the first thing I doe: and Make
1542 his *Inuocation* in the *Name* of others.
1546 *Thou, whose gentle Forme, and Face,*
fill'd latelie this Egiptique Glas:
By th'emperious powrefull Name
and the Vniuersall Fame
1550 *of the Mightie-Black-house-Queene*
I Coniure Thee to be seene
What? see you nothing yet?
1553 *wh.* Q^s. *P.* Not any part.
\+ 'pray try an other:
\+ B^l. Q^s. *P.* You shall haue your Will.
1554 *I double my Commaund and Powre,*
and at the instant of this howre.
Inuoake Thee in the white-Queenes Name
with stay: for Time, and Shape the same.
What see you yet?
wh. Q^s. *P.* There's nothing showes at all.
1560 B^l. Q^s. *P.* My Truth reflects the cleerer: Then now fix
and bles your faire eie, with your owne for ever.
Thou well-Compos'd, by Fates-hand drawne
to enioy the White-Queenes-Pawne

of

of whom Thou shalt (by Vertue mett)
many gracefull Issues gett.
By the Beutie of her Fame
By the whitenes of her Name
By her faire, and fruitfull Loue,
By her Truth, (that mates the Doue,)
1570 *By the Meekenes of her Mind*
By the Softnes of her Kind,
By the Lustre of her Grace.
By all Theis thou art summond to this Place.

Musique:
The B^{l}. B^{ps}. Pawne (as in an Apparition) Comes richely habited.

Hark how the *Aire*, enchaunted with your *Praises*,
1575 and *His Approach*; those *words*, to sweet *Notes* raises.
1580 *wh. Q^{s}. P.* Oh, let him stay a While: a litle Longer.
1582 Yf He be Mine, why should he part so soone?
1583–4 *B^{l}. Q^{s}. P.* why this is but the Shadow of yours: How doe you?
1585 *wh. Q^{s}. P.* oh, I did ill to give consent to see It:
What Certentie is in our Blood, or State?
What we still wryte, is blotted out by *Fate*.
Our Wills are like a *Cause*, that is *Law*-tost.
What one *Court orders*, Is by an other Crost.
1590 *B^{l}. Q^{s}. P.* I find no fit place, for this *Passion* here
'tis meerely an Intrudo^{r}; He is a Gentleman
most wishfully Composd: Honor growes on him,
and Wealth pilde-vp for him: 'hath Youth enough too:

And

941 *find*] *d* altered

And yet, in the Sobrietie of his Countenaunce
1595/6 Grave as a *Tetrach*: Where's the Emptines?
What can you more request?
wh. Q^{s}. *P*. I doe not knowe
What Answeare yet to make: It doth require
1600 a Meeting 'twixt my *Feare*, and my *Desire*.
\+ B^{l}. Q^{s}. *P*. She's caught: and (which is strange) by her most Wronger.
1601 – *Exeunt*.|

1063

Actus Quartus.

\+

Sce^{a}. prima.

1604–5/1647–8 ***The Black B^{ps}. Pawne: & y^{e} white, & Bl: Queenes Pawnes.***

1645 B^{l}. B^{s}. *P*. **Yonder's** my *Game*, Which (like a Politique *Ches-master*)
1646 I must not seeme to see.
1648 *wh*. Q^{s}. *P*. oh, my hart: 'tis He:
1650 the verie self-same that the *Magicall Mirrour*
presented lately to Me.
B^{l}. Q^{s}. *P*. and how like.
a most regardles Stranger he walkes by
meerely Ignorant of his *Fate*? you are not minded.

the

39.

(the principallst part of him:) What strange Misteries
Inscrutable *Loue* works by.
wh. Q^s. *P*. The Time you see
is not yet Come.
1659 *B*^{l}. Q^s. *P*. but 'tis in our powre now
1661 to make it obserue vs, and not we it.
wh. Q^s. *P*. I would force nothing from It's proper Vertue.
Let Time haue his Full Course: I had rather die
the modest death of vndiscouerd *Loue*,
then haue Heavens least, and lowest Servant suffer,
1666 or in his Motion receive Check for Me.
1670 *B*^{l}. Q^s. *P*. he comes this way agen.
wh. Q^s. *P*. oh, there's a Traito^r
leapd from my hart, into my Cheeke already
that will betray all to his powrefull eie
if it but glaunce vpon Me.
B^l. Q^s. *P*. by my Veritie,
1676 looke, He's past by agen drownd in Neglect.
1681 'twer pitty he should dwell in Ignorance longer.
1686 Absolute-Sir) with your most noble Pardon
for this my rude Intrusion, I am bold
to bring the knowledge of a Secreat neerer
by many Daies (Sir) then it would arive
1690 in Its owne proper Revelation with you.

pray

975 Q^s.] *Q* altered

'pray turne and fix: Do you know yond Noble Goodnes?
B^l. B^s. *P* 'tis the first Mynut, mine eie blessd me with her.
and cleerely showes how much my knowledge wanted
not knowing her till now.
1695 B^l. Q^s. *P*. She's to be likd then?
1709 'pray mark her once agen, then follow me
and I will show you her, must be your Wife (Sir.)
B^l. B^s. *P*. The Mistery extends, or els Creation
hath set that Admirable Peece before vs,
to Choose our Chast Delightes by.
B^l. Q^s. *P*. 'please you follow (Sir)?
B^l. B^s. *P*. What Art haue you, to put me on an Obiect
1716–17 and cannot get me off? 'tis paine to part from't – *Ex*^{t}.
1718 *wh*. Q^s. *P*. Yf there prove no Check in that *Magicall Glas* now.,
but my *Proportion* come as Faire, and full
1720 into his eye, as His into mine Lately.
1721–2 then I am confirmd, he is mine owne for ever. {– *Enter agen.*
1723 B^l. B^s. *P*. The very self same that the *Mirro*^{r} blest me with,
from Head to Foote, the Beutie, and the Habit;
Kept you this Place still? Did you not Remove (*Ladie*?)
wh. Q^s. *P*. Not a Foote Farther (Sir)
B^l. B^s. *P*. Is't possible?
I would haue sworne, I had seene the Substance yonder

'twas

989 *'tis*] *s* altered 1010 *would*] *ld* altered

'twas to that *Lustre*, to that life presented.
1730 *wh*. Q^{s}. *P*. even so was yours to Me (Sir)
B^{l}. Q^{s}. *P*. saw you mine?
wh. Q^{s}. *P*. perfectly cleere: No sooner my *Name* vsd
but yours appeerd.
B^{l}. B^{s}. *P*. iust so did yours at mine, now.
1735 B^{l}. Q^{s}. *P*. Why stand you Idle? will you let Time Cozon You
1740 Can you be more then *Man*, and *wife* assignd?
1741 and by a Powre, the most Irrevocable?
1746 B^{l}. B^{s}. *P*. She speakes but Truth, in this: I see no reason then
that we should misse the Rellish of this Night,
but that we are both shame-facd.
wh. Q^{s}. *P*. how? this Night (Sir?)
1750 Did not I know you must be Mine (and therein
Your Priviledge runs strong) for that loose Motion
You never should be. Is it not my Fortune
to Match with a pure Mind? then am I miserable.
The *Doues*, and all Chast loving-Winged Creatures
haue their *Paires* Fit, their *Desires* iustly mated,
Is Woman more vnfortunate? a Virgin?
(the *May of Woman*?) *Fate* that hath ordaind (Sir)
We should be *Man* and *wife*, hath not given warrant
for any Act of *Knowledge*, till we are soe.

Tender

42.

1760 *B^{l}. B^{s}. P.* Tender-eid-Modestie, how it gives at this?
(I am as far-off (for all this strange Imposture)
as at First Enter-View: Where lyes our *Game* now?
You know I cannot *Marrie* by mine *Order*?
B^{l}. Q^{s}. P. I know you cannot (Sir) Yet you may Venture
1765/1767 vpon a *Contract*? Surely you may Sir,
1768 without all question so Far, without danger.
1769 or any Staine to Your Vow, and that may Take her.
1772 *B^{l}. B^{s}. P.* Be not so lavish of that Blessed Spring.
You haue wasted that vpon a cold occasion now
would wash a sinfull Soule, white: By our Love-Ioies
that Motion shall nere light vpon my Tongue more,
till we are *Contracted*? Then I hope you'are Mine?
wh. Q^{s}. P. In all iust dutie ever.
B^{l}. Q^{s}. P. Then? doe you question it?
pish: then you'are *Man*, and *Wife*, all but *Church-Ceremony*:
1780 'pray Let's see that don First, She shall doe reason then.
(Now I'll enioy the Sport, and Cozon 'em both:
1782–3 My Bloods-*Game*, is the Wages I haue wrought For – *Exeunt*.

+ ## ***Sce^{a}. secunda.***

1784 ***The Bl. Knight, & his Pawne: Then y^{e} Fat-Bishop:***
\+ ***Then the Black-king.***

1785 *B^{l}. K^{t}.* (*Pawne*, I haue spoak to the *Fat-Bishop* For Thee,

I'll

1052 *Exeunt*] *t* not crossed

I'll get thee *Absolution* from his owne Mouth.
Reach me my *Chaire of Ease*, my *Chaire of Cozonage*,
Seauen-thousand Pound in Women, Reach me that.
I live-of-life, to sit vpon a Banck
1790 of *Hereticque* Gold; (oh soft, and gently Sirha)
There's a fowle Flaw i'th' Bottom of my *Drom* (*Pawne*)
I nere shall make sound Soldier, but sound Treacher
with any He, in *Europe*: How now (*Qualme*?)
Thou hast the pukinst Soule that ere I mett with
it cannot beare one Suckling Villany:
Mine can digest a Monster, without Cruditie,
a Sin, as weightie as an Ellephant.
and never wamble For't.
1799 *Paw*. you haue bin vsd to't (Sir)
1803 Yf I had got seaven thousand pound by *Offices*
1804–5 and gulld downe that: the Bore would haue byn bigger.
1809 B^{l}. K^{t}. Hadst thou betrayd the *white-house* to the *Black*
1810 Beggard a *Kingdome* by *Dissimulation*:
Vnioynted the faire Frame of *Peace*, and *Traffique*:
Poysond *Allegeance*; Set *Faith* back, and wrought
Womens soft Soules (even vp to Masculine Malice)
to pursue Truth to death, yf the *Cause* rowzd 'em.
that Stares, and Parrots, are First taught to Cursse thee.
1816 *Paw*. I mary (Sir) heere's swapping Sins indeed.
1828 B^{l}. K^{t}. Sirha: I haue sould the *Groome* o'th' *Stoole* six times,

and

1060 *live*] sic, read *love* 1062 *There's*] *T* altered from *S* 1073 K^{t}.] full-stop extended horizontally, but not a true hyphen

and receiu'd Money of six seuerall *Ladies*
Ambitious to take place of *Baronetts* wives.
To three old *Mummey-Matrons*, I haue promisd
the *Mother-ship* o'th' *Maides*: I haue taught o^{ur} *Frends* too
to Convay *White-house Gold* to our *Black-kingdome*,
in Cold Bakde-*Pasties*, and so Cozon *Searchers*.
For Venting *hallowed Oile*, *Beades*, *Medals*, *Pardons*,
Pictures, *Veronica's Heads*, in Privat *Presses*,
that's don by one in th'abit of a *Pedler*.
Letters convaid in *Rowles*, *Tobacco Balles*,
When a *Restraint* comes, by my politique Councell
some of our *Iesuites* turne *Gentlemen-Vshers*,
some *Faulconers*, some *Park-keepers*; and some *Huntes-men*:
One took the Shape of an old *Ladies Cooke* once,
and dispatchd Two Chewres on a Sonday Morning
(the *Altar*, and the *Dresser*) I'pray what vse
put I my *Som̃er Recreation* to,
but more t'enforme my *knowledge* in the State
and *Strength* of the *white-Kingdome*? No *Fortification*,
Hauen, *Creeke*, *Landing-Place* 'bout the *white-Coast*
but I got *draught*, and *Platforme*: Learnd the depth
of all their *Channells*, knowledge of all *Sands*,
Shelves, Rocks, and *Riuers*, for Invasion properst.
A Catolougue of all the *Nauy Roiall*:
The *Burthen* of each *Ship*: the *Brassy Murderers*,

The

1091 *Balles,*] followed by pen-rest   1092 *by*] altered   1096 *Chewres*] oblique pencil-stroke through *h*

The *Nomber of the Men*, to What *Cape* bound.
Agen, (for the *Discouerie* of the *In-Lands*)
Never a *Sheire*, but the State better knowne
to Me, then to her Brest *Inhabitants.*
What *Powres of Men*, and *horse*, *Gentries-Reuenewes*;
Who well-affected to our State, who Ill:
1860 Who, neither Well, nor Ill: All the *Neutralitie.*
Thirtie eight thousand Soules, haue byn seducd (*Pawne*)
since the *Gaoles* vomited with the *Pill* I gaue'em.
Paw. sure you put *oile* of Toad, into that phisique (Sir)
B^l. K^t. I am now about a Master-peece of *Play*,
To Entrap the *white-knight*, and with falce Allurements.
Entice him to the *Black-House*: more will follow.
Whilst our Fat-*Bishop* setts Vpon the *Queene*
Then will our *Game* lye sweetely.
1869 *Paw. He's* come now (Sir)
1871 *fat B.* Heere's *Taxa Pœnitentiaria* (*Knight*)
The *Booke of Generall Pardons* of all *Prices.*
I haue byn searching for his Sin, this half howre,
and cannot light vpon't.
B^l. K^t. that's strange: Let me see't.
Paw. (*Pawne* wretched that I am) hath my *Rage* don that
there is no *President of Pardon* for?
8– { *B^l. K^t. For willfull Murder: Thirteene Pound, fowre shillings, and*
0 *six-pence.* (that's reasonable cheape) *For Killing. Killing: Killing*

why

46.

1881 why heere's nothing but *Killing* (*Bishop*) on this side?
fat B. Turne the Sheete over, you shall find *Adulterie*
and other Triviall Sins.
B^{l}. K^{t}. *Adulterie*? (oh, I am in't now) *For Adulterie*
a cople of shillings: And for Fornication
fiue Pence. (Theis are two good penney-Worthes: I cannot see
1887–1890 how a Man can mend himself.) *For lying with Mother*
Sister, and Daughter (I mary Sir) *Thirteene Pound, Three*
shillings, three Pence. The Sins Gradation right: paid all in *Threes*
1891 *fat B*. you haue read the *Storie* of that *Monster* (Sir)
that got his *Daughter, Sister, and his wife*,
of his owne *Mother*?
B^{l}. K^{t}. *Symonie: nine-Pound.*
fat B They may thanck me for that: 'twas Nineteene
before I came: I haue mittigated many of the Som̃es.
B^{l}. K^{t}. *Sodomie; six Pence:* (you should put that Som̃e
ever on the Backside of your Booke (*Bishop*)
fat B. There's fewe on's very forward (Sir)
1900–1 B^{l}. K^{t}. what's here (Sir?) *Two old Presidents of Encouragement?*
fat B. I: those are *auncient Notes.*
1903–5 B^{l}. K^{t} *Giuen as a Gratuitie, for ye Killing of an Hereticall Prince,*
with a Poisond knife; Duccats fiue Thousand.
1906 *fat B*. True (Sir) that was paid.
B^{l}. K^{t}. *Promis'd also to Docto^{r} Lopez, for Poysoning ye Maiden-*
Queene

Queene of the white-Kingdome, Duccats Twentie-thousand.
which said some was afterwards giuen as a Meritorious Almes
to ye Nunnerie at Lisbon, hauing at this present Ten
12–13 *thousand Pound more at Vse in ye Towne-house of Antwerpe.*
1914 *Paw.* What's all this to my Conscience? (worthie *Holines*)
I sue for *Pardon*; I haue brought Money with Me.
fat B. You must depart; You see there is no *President*
of any *Price*, or *Pardon* for your *Fact*.
Paw. most miserable: are fowler Sins remitted?
Killing? nay willfull *Murder*?
1920 *fat B.* True: there's Instance:
Were you to kill him: I would pardon You:
ther's president for that, and *Price* set downe,
but none for *Guelding*.
Paw. I haue pickd-out Vnderstanding now for ever
out of that *Cabalistique* Bloody *Ridle*.
1926 I'll make away all my *Estate*, and *Kill him*,
\+ and by that *Act*, obtaine full *Absolution*. – *Exit*
1928 B^{l}. *K.* Why *Bishop*: *Knight*, wher's your *Remoues*? you^{r} *Traps*?
Stand you now Idle, in the *heat of Game*?
1930 B^{l}. *K^{t}*. My life for yours (*Black-Soueraigne*) *the Game's ours.*
I haue wrought vnder-hand for the *white-Knight*,
and his Brave *Duke*, and find 'em Com̄ing both.

Then

48.

fat B. Then for their Sanctimonious *Queenes* Surprizall
in that State-puzzell, and distracted hurrey,
trust my Arch Subtletie with.
B^{l}. *K*. oh Eagle-Pride,
Never was *Game* more hopefull of our *Side*.
B^{l}. *K*^{t}. Yf *Bishop Bull-Beof*, be not Snap'd at next Bowt,
1939–40 (as the *Men* stand) I'll neuer trust *Art* more – *Exeunt*/

+

Sce^{a}. *Tertia.*

1941–7 ***A Domb showe*** ***The Bl: Queenes Pawne (with a Taper) Conducts y^{e} wh. Queenes-Pawne (in her night attire) into one Chamber; Then Conuaies y^{e} Bl. B^{ps}. Pawne, into an other Chamber, So puts out y^{e} Light, & followes him.***

+

Sce^{a}. *Quarta.*

1948 ***The white-Knight, & wh. Duke, y^{en} y^{e} Bl. Knight: Then***
\+ ***y^{e} white-Queene: y^{e} Fat Bishop: y^{e} white Bishop; &***
\+ ***y^{e} wh. King.***

1949 *wh. K*^{t}. True *Noble Duke*, Faire *Vertues* mostEndeer'd-one,
1950 Let vs prevent their Ranck Insinuation
With Truth of *Cause*, and *Courage*; Meete their *Plotts*
with Confident *Goodnes*, that shall strike 'em groveling.

Sir

1184 *Beof*] sic 1187 *Queenes*] ^{2}*e* altered from *a* 1192 ^{2}*Knight*:] *t* not crossed

49.

wh. D. (Sir) All the *Iyns*, *Traps*, and Alluring Snares
the Devill hath byn at work, since eightie-eight on,
are layd for the Great hope of this *Game* onely.
wh. K^{t}. why, the More Noble will *Truthes-Triumph* be,
When they haue woond about our Constant Courages
the Glittringst Serpent, that ere *Falsehood* fashiond,
and, glorying most in his resplendant *Poisons*,
1960 Iust heaven can find a Bolt, to bruize his head.
wh. D. Looke, would you see *Distruction* lye a Suñing?
In yonder Smile, sitts Blood, and Treacherie basking.
In that perfidious Modell of *Face-Falsehood*
Hell is drawne gryñing.
wh. K^{t}. what a paine it is
for *Truth* to Faigne a litle?
B^{l}. *K*^{t}. oh Faire *Knight*;
The *Rising Glorie* of that *House* of *Candor*,
haue I so many *Protestations* lost?
1970 lost? lost? quight lost? Am I not wo^{r}th your Confidence?
I that haue vowd the *Faculties* of Soule,
1972 Life, Spirit, and Brayne, to Your sweet *Game* of Youth?
1978 How often haue I changd for your Delight
the *Roiall Presentation* of my *Place*
80/1982 into a Mymick-*Iester*? a Meere wanton?
1983 I'll tell you what I told a *Sauoy Dame* once,

New

1200 *at*] *t* not crossed

New Wed, high, Plump, and lusting for an *Issue*,
Within the Yeere, I promisd her a Child
if She could stride ouer St *Rumbants Breeches*
(a *Relique* kept at *Mechlin*) The next Morning
one of my *Followers* old *Hose*, was Convaid
into her *Chamber*, Where She tryde the Feate,
1990 By that, and a *Court-Frend*, after grew Great.
wh. K^{t}. why who could be without Thee?
B^{l}. *K*^{t}. I will Change
to any Shape, to please you: and my Ayme,
hath byn to Wyn Your Love, in all this *Game*.
wh. K^{t}. Thou hast it nobely, and We long to See
the *Black-house Pleasure*, *State*, and *Dignitie*.
B^{l}. *K*^{t}. of *Hono*^{r} you'll so Surfeyt, and Delight
1998–9 You'll nere desire agen, to see the *white* – *Ex*^{t}.
+ *wh. Queene*. My *Loue*, my *Hope*: my *Deerest*, oh, he's gon:
Ensnard, Entrap'd, Surprizd amongst the *Black-ones*,
I never felt Extremitie like this.
I shalbe Taken: The *Game's* lost: I am set vpon:
Oh 'tis the *Turne-Coate-Bishop* (having watch'd
th'advantage of his *Play*,) Comes now to Ceize on Me
oh, I am hard Be-set, distresd most miserably.
fat. B. 'tis Vaine to stirr. Remove wch way you can
I take you now: This is the *Time* we haue hopd For.

Queene

1226 *The*] *T* blotted 1238–1288] not in Trinity

(*Queene* You must downe: there is no remedie
The *Black-Kings* Blood, burnes for thy *Prostitution*,
and nothing but the Spring of thy Chast Vertue
can Coole his *Inflamation*: Instantly
he dies vpon a *Pluresie of Luxurie*
Yf he *de-flowre* thee not.
wh. Q. oh streyt of misery.
wh. B. And is your *Holines* his Devine *Procuro*^{r}?
fat B. The Devill's in't: I am Taken by a *Ring-Doue*:
Where stood this *Bishop*, that I saw him not?
wh. B. You were so Ambitious you lookd over Me:
You aym'd at no lesse *Person* then the *Queene*,
(the Glory of the *Game*) yf She were won
the way were open to the *Master-Check*.
Which (looke you) He, and His live to give you.
Honor, and *Vertue* guid him in his Station.
wh. Q. oh, my saffe Sanctuarie:
wh. K. Let heavens Blessings
be mine no longer then I am thy Sure-One.
The *Doues-house* is not saffer in the *Rock*
then Thou in my Firme Bosome.
wh. Q^{s}. I am blessd in't.
wh. K. Bishop, Thou hast don our *White-house* gratious Service,

and

1268 Q^{s}.] sic, read *Q*.

and worthie the faire *Reuerence* of thy *Place*.
ffor Thee (*Black-holines*) that workst out thy Death
as the *Blind-Moale*: 'twer well indeed with Thee,
yf like that Verminous Laboure^{r}. (which thou imytatst
in Hills of *Pride*, and *Malice*) when Death puts thee Vp,
the silent Grave, might prove Thy *Bag* for ever,
(no deeper *Pit* then that) ffor thy Vaine hope
of the *White-knight*, and his most firme *Assistant*
Two *Princely Peeces*, wch I know thy Thoughtes
give lost for ever now: My strong Assuraunce
of their fixd Vertues, could you let in Seas
of populous Vntruthes against that *Fort*
'twould burst the prowdest Billowes

wh. Q. my Feare's past then.

wh. K. Take that *Prize* hence: Goe *Reuerend of Men*,
Put Couetousnes into the Bag agen.

fat B. The *Bag* had need be strong, or it goes to Wrack,
Sin, and my *waight*, will make a strong one Crack.

– *Exeunt*|

1273 *imytatst*] *i* inserted afterwards   1278 *Two*] oblique pen-stroke through *T*

2001 *Actus Quintus.*

+ *Sce^{a}. prima.*

002–3 ***The Black-Knight (in his Litto^{r}) & ye Bl: B^{ps}. Pawne***
+ ***aboue: Then ye Black-house, (meeting the white***
+ ***Knight, and white Duke.***

2005 *B^{l}. K^{t}* ***Is*** the *Black-Bishop's Pawne*, the *Iesuite* planted above for his Concise *Oration*?

B^{l}. B^{s}. P Ecce Triumphanti, Me fixum Cæsaris Arce.

B^{l}. K^{t}. Ar't there (my *holy Boy*?) Sirha: *Bishop Tumbrell* is Snap'd i'th' *Bag* by this time.

2010 *B^{l}. B^{s}. P. Hæretici pereant sic.*

B^{l}. K^{t}. All lattin? sure the *Oration* hath infected him.
2012 Away: make haste; they are Coming.

17–27 *B^{l}. B^{s}. P. Si quid mortalibus Vnquam Oculis hilarem, et gratum aperuit Diem, Si quid per amantibus Amicorum Animis Gaudium attulit, peperituè Lætitiam (Eques Candidissime-prælucentissime) fælicem profectò Tuum, a Domo Candoris, ad Domum Nigritudinis Accessum, promisisse, peperisse, attulisse fatemur. Omnes Aduentus Tui conflagrantissimi, Omni qua possumus Lætitia, Gaudio, Congratulatione, Acclamatione, Animis obseruantissimis*

Affec=

Affectibus diuotissimis, Obsequijs Venerabundis Te
Sospitem Congratulamur.

2028 B^{l}. *K.* Sir,) In this short *Congratulatorie Speeche*
You may Conceive how the whole *House* affects you.
2030 B^{l}. K^{t}. The *Colledges*, and *Sanctimonious Seede-Plotts.*
\+ *wh.* K^{t}. 'tis cleere: and so acknowledgd (*Roiall* Sir)
2031 B^{l}. K^{t}. Harck: (to Enlarge your *Welcom*) from all *Parts*
is heard sweet-sounding *Aires*: *Abstruse-Things* open,
of voluntary freenes: and yond *Altar*
(the *Seate* of *Adoration*) seemes t'adore
the *Vertues* you bring with you.

{ ***An Altar*** discovered with ***Tapors*** on it: and ***Images*** about it

wh. K^{t}. There's a Taste
2037 of the *old vessell* still.
wh. D. th'*Erronious Rellish.*

2040 ***Song.***

***Wonder** work some strange-Delight*
*(this **Place** was neuer yet with out)*
*To welcom the faire **white-House-Knight,***
and to bring our Hopes about.
May from y^{e} ***Altar, Flames** aspire,*
*Those **Tapers,** set themselues on fire:*
May senceles Things our Ioies approue,
2048 *and those **Brazen-Statues** moue*
\+

{ *The **Images** moue in a Dance* *quickend*

1313 *Plotts.*] followed by pen-rest

55·

2049 *quickend by some Powre aboue;*
2050 *or what more strange to show our Loue.*
\+ B^{l}. K^{t}. A happie *Omen* waytes vpon this howre:
\+ All *Moue* (portentously) the *Right-hand* way
51–2 B^{l}. *K*. Come, let's sett free, all the most Choice Delightes
53–4 that ever adornd Daies, or quickend Nightes – *Exeunt*/

\+ **Sce^{a}. secunda:**

2055 ***The white Queenes Pawne, & Black-Bishop's Pawne.***
\+ ***Then y^{e} Black Queenes-Pawne: Then y^{e} white***
\+ ***Bishop's Pawne, & y^{e} Bl. Knights Pawne.***

2056 *wh*. Q^{s}. *P*. I see 'twas but a Triall of my Dutie now,
'hath a more modest Mind, and in that Vertue
2058 most worthelie hath *Fate* provided for Me:
2060 Hah? ''tis the Bad man, in the *Reuerend habit*
61/+ Dares he be seene agen (Traito^{r} to Holynes)
2062 oh Marble fronted Impudence, and knowes
063–4 how Ill 'hath vsd me? I'am ashamd he blushes not.
2065 B^{l}. B^{s}. *P*. Are you yet stoard with any Womans pittie?
Are you the *Mistris* of so much Devotion,
Kindnes, and Charitie, as to bestow

an

1346 *''tis*] double apostrophe   1351 *the*] *h* touched up

2068 an Almes of *Loue*, on Your poore *Suffrer*, yet?
2070 *wh. Q*^{s}*. P.* Sir, for the *Reuerend Respect* you ought
to give to Sanctitie (though none to Me)
in being her *Seruant* vowd, and weare her Livory:
Yf I might Councell you, you should nere speake
the Language of Vnchastnes in that *Habit*,
You would not thinck how ill it doth with you.
The world's a Stage, on Which all *Parts* are plaid,
you'lld thinck it most Absurd to see a Devill
presented there, not in a Devills shape,
or (Wanting one) to send him out in yours,
2080 you'ld raile at that for an Absurdetie
no *Colledge* ere Com̃itted: for Decorum-sake, then,
for Pitties cause: for sacred Vertues hono^{r},
yf you'll persist still in your Devills-Part
present him, as you should doe: and Let one
that Carries vp the Goodnes of the *Play*
2086 Come in that Habit, and I'll speake with him.
2094 Is there so litle hope of you, to smile (Sir)?
2095–6 *B*^{l}*. B*^{s}*. P.* yes: at Your Feares: at th' Ignorance of your Powre,
2097 the litle vse you make of Time, youth, *Fortune*,
Knowing you haue a *Husband* for *Lusts* shelter,
you dare not yet make bold with a *Frends* Comfort.

this

1369 *of*] *f* touched up

This is the plague of Weakenes.
wh. Q^s. P. so hot burning
the Sillables of Sin fly from his Lipps
as if the Letter came new-cast from Hell.
B^l. B^s. P. Well: setting a-side the Dish you loath so much
(which hath byn hartely tasted by your Betters,
I Come to *Marrie* you to the *Gentleman*
that last enioyd you: I hope that pleases you?
there's no im̃odest Rellish in that office?
wh. Q^s. P. strange, of all Men, he should first Light on him
to Tye that holy *Knot*, that sought t'vndooe me?
Were you requested to performe that Busynes (Sir?)
B^l. B^s. P. I name you a sure Token.
wh. Q^s. P. As for that (Sir)
Now y'ar most wellcom: and my faire hope's of yo^u
you'll never break that sacred *Knot* you tye once
with any Lewd Solliciting hereafter.
B^l. B^s. P. But all the Crafte's in getting of it knit:
You are all one fire to make your Cozoning Market:
I am the *Marrier*, and the *Man*; doe you know me?
Doe you know me (nyce-*Iniquitie*, Strict *Luxurie*,
and holy *whoredome*,) that would clap on *Marriage*
with all hot speed to soalder vp your *Game*?

See

58.

See what a Scourge *Fate* hath provided for Thee.
You were a *Maid*, sweare still: y'ar no worsse now
I left you as I found you: haue I startled you?
I am quitt with you now for my Discovery,
2127–8 Your Out-cries, and your Cuñings: farewell *Broccadge*.
wh. Q^{s}. P. Nay, stay, and heare me but give thancks a litle
(yf your eare can endure a Work so gratious)
then you may take your pleasure.
B^{l}. B^{s}. P. I haue don that.
wh. Q^{s}. P. That Powre, that hath preseru'd me from this Devill
B^{l}. B^{s}. P. how?
wh. Q^{s}. P. This: that may Challenge the *Cheif Chaire* in *hell*
and sit above his *Master*?
B^{l}. B^{s}. P. Bring in *Merit*?
wh. Q^{s}. P. That sufferdst him, (through Blind *Lust*) to be ledd
last night, to the *Action* of some *Com̃on-Bed*.
B^{l}. Q^{s}. P within}. Not over-Com̃on neither.
B^{l}. B^{s}. P. hah? what Voice is that?
wh. Q^{s}. P. of *Virgins* be thou ever hono^{r}ed:
You may goe now: you heare I haue given Thancks Sir)
B^{l}. B^{s}. P. Heere's a strange *Game*: Did not I lye with you?
within}. Noe:
B^{l}. B^{s}. P. what a Devill art thou?

I

1399 *Scourge*] followed by mark like hyphen, probably accidental   1410 *hell*] followed by pen-rest

wh. Q^{s}. *P*. I will not answeare you (Sir)
after-Thanckes-giving.
B^{l}. B^{s}. *P*. why, you made promise to Me
)/2153 after the *Contract*? And you were prepar'd for't?
and set your Ioies more high?
within}. Then you could reach (Sir)
2156 B^{l}. B^{s}. *P*. This is a Bawdy *Pawne*; I'll slytt the throat on't.
2158 B^{l}. Q^{s}. *P*. What? offer Violence to your Bedfellow?
to one that workes so kindly, without *Rape*?
2160 B^{l}. B^{s}. *P*. my Bed-Fellow?
B^{l}. Q^{s}. *P*. Doe you plant your Scorne against me?
Why, when I was *Probationer* at *Bruxells*,
that Engine was not knowne. Then *Adoration*
filld-vp the Place; and wonder was in Fashion:
Is't turnd to the Wild Seed of Contempt so soone?
6/2167 Can five yeeres stamp a *Bawd*? It is no longer (Sir)
since you were Cheif Agent for the *Transportation*
of *Ladies* Daughters, yf you be remembred;
2170 Some of their *Portions* I could name; Who pursd 'em too:
They were soone disposesd of worldly Cares
that came into your Fingers.
B^{l}. B^{s}. *P*. shall I heare her?
2174 B^{l}. Q^{s}. *P*. Holy Derision, yes: till thine Eare swell:
2176 Whose [Child] Neice was she, you poysond with Child twice,

and

1438 *since*] followed by mark resembling colon erased, not deleted, but legible 1441 *worldly*] ^{2}l touched up 1445 [*Child*]]

60.

and gave her out possessd with a fowle Spirit
When 'twas indeed your *Bastard*?

2179–80 *B^{l}. B^{s}. P* I am Taken
in mine owne Toiles.

wh. B^{s}. P. yes: and 'tis Iust you should be.

wh. Q^{s}. P. And thou (lewd *Pawne*) the shame of Womanhood

B^{l}. B^{s}. P I am lost of all hands.

B^{l}. Q^{s}. P. And I cannot feele
the waight of my *Perdition*, now He's taken.
it hath not the Burthen of a Grashopper.

2188 *B^{l}. B^{s}. P.* Thou whore of *order*, *Cockatrix in voto*.

2190 *B^{l}. K^{ts}. P.* yond's the *white-Bishop's Pawne*: I'll *Play* at's Hart now.

wh. Q^{s}. P. how now (*Black-Villaine*) wouldst thou heape a Murder
on thy first fowle offence? oh Merciles Blood-hound
'tis time that thou wert Taken.

B^{l}. K^{ts}. P. how? prevented?

wh. Q^{s}. P. for thy sake, and that *Partner* in thy *Shame*
2196–7 *I'll neuer know Man farther then by Name. – Exeunt.*|

+

Sce^{a}. Tertia.

2198–2200 ***The Black-House, & ye white-knight, & Duke: Then the***
+ ***white King, Queene, Bishop.***

2201 *wh. K^{t}.* You haue Enrich'd my *Knowledge* (*Roiall* Sir.)

and

1466 *Bishop*.] followed by pen-rest

61.

and my Content togeather.
B^{l}. *K*. 'stead of *Riot*,
we set you onely welcom: Surfeyt is
a Thing that's seldom heard of in theis *Parts*,
wh. *K*^{t}. I heare of the more Vertue when I misse on't.
B^{l}.. *K*^{t}. We doe not vse to Bury in o^{ur} Bellies
Two hundred Thousand Duccatts, and then boast on't
Or Exercise th'old *Romaine* painefull *Idlenes*
2210 with Care of Fetching *Fishes* far from home,
The *Golden-headed Corasine*, out of *Egipt*,
the *Salpa* from *Eleusis*, or the *Pelamis*
2213 (which some Call *Som̃er-whiting*) from *Calcedon*:
2217 We Cramb no *Birds*, nor (Epicurean-like)
2218 enclose some *Creekes* o'th' Sea, as *Sergius Crata* did.
2223 Nor doe we Imytate that Arch-Gurmondizer
with Two and Twentie *Cowrses* at one Dinner,
and betwixt every *Cowrse*, He, and his *Guests*,
2226–7 washd, and vsd *Woemen*,; Then sat downe, and strengthend
2228 Lust, swym̃ing in their Dishes: Which no sooner
was tasted, but was ready to be vented
2230 *wh*. *K*^{t}. most impious Epicures.
B^{l}. *K*^{t}. we Com̃end rather
(of Two Extreemes) the *Parsimony* of *Pertinax*,
who had half Lettysses set vp to serve agen:

Or

1474 *Duccatts*] ^{1}*t* in tail of *y* above 1475 *Exercise*] *E* altered 1479 *whiting*] could be read as *whitnig* (see Introduction, p. 7) 1481 *Sea*] *a* touched up

62.

Or his Successo^{r} *Iulian*, that would make
Three meales of a Leane-Hare, and often Sup
with a Greene *Fig*, and wipe his Beard, as we can:
The old Bewaylers of Excesse in those daies
Complaind there was more Coyne bid for a Cooke
then for a war-horse: But now Cookes are purchased
2240 after the rate of Triumphes: and some Dishes
after the rate of *Cookes*, Which must needes make
some of Your *white-house Gurmondizers* (specially
your Wealthie fat *Plebeians*) like the *Hogs*
2224 which *Scaliger* Cytes, that could not Move for fat
2257 *wh. K^{t}*. Well: y'ar as sound a Spoakes-man (Sir) for *Parsimony*
Cleane Abstinence, and scarce one Meale a-day
2259 as ever spake with tongue.
2262 *B^{l}. K*. hee'll raise of any thing.
2263 *wh. K^{t}*. I shalbe half a fraid to ffeed hereafter.
2270 *B^{l}. K^{t}*. how you mis-prize? This is not meant to You-ward.
You, that are woond vp to the height of Feeding
by Clyme, and *Custome*, are dispensd with all
2273 You may eate *Kid*, *Cabrito*, *Calf*, and *Tons*,
2275 Nay, the Franckd-hen, (fattend with Milk, and Corne.
2278 *wh. K^{t}*. Well: for the *Food*, I am happely resolu'd in,
but for the *Diet* of my *Disposition*,

there

63.

there Comes a Troble: You will hardly find
Food, to please that.
B^l. K^t. It must be a strange *Nature*
We cannot find a Dish for: having *Policie*
(the *Master-Cooke of Christendom*) to dresse it:
'pray name your *Natures Diet*.
wh. K^t. The First *Messe*
is hot *Ambition*
B^l. K^t. That's but seru'd in *Puff-Paste*.
Alas, the meanest of our *Cardinalls-Cookes*
can dresse that *Dinner*: Your *Ambition* (Sir)
can fetch no Farther Compas then the World?
wh. K^t. that's certaine (Sir)
B^l. K^t. We are about that alredy.
And in the Large *Feast* of our Vast *Ambition*
We Count but the *White-Kingdome* (whence You came from)
the *Garden* for our *Cooke*, to pick his *Sallads*:
The *Food's* Leane *France*, Larded with *Germanie*:
Before which Comes the *Graue* Chaste *Signorie*
of *Venice*, seru'd in (Capon-like) in white-broth.
From our Cheif *Ouen*, *Italie*, the *Bake-meates*,
Sauoy, the *Salt*, *Geneua* the Chip'd *Mantchett*.
Below the *Salt*, the *Netherlands* are placd,
(a *Com̃on*-Dish, at'Lower-end o'th, *Table*

for

for Meaner *Pride* to fall to) ffor *our Second Cowrse*
a Spit of *Portugalls* (seru'd in for *Plouers*,)
Indians, and *Moores* for *Black-Birds*: All this while
Holland stands ready melted, to make Sawce,
on all occasions: When the Voider comes:
And with such Cheere, our full hopes we suffice,
Zealand saies *Grace* (for fashion) then we rise.
wh. K^{t}. Heere's Meat enough (on-Conscience) for *Ambition*
B^{l}. K^{t}. Yf there be any Want There's *Switzerland*,
Polonia, and such *Pickeld-Things* will serve
to furnish out the Table.
wh. K^{t}. you say well (Sir)
But heere's the misery; When I haue stopd the mouth
of one *Vice*, there's an other gapes for *Food*:
I am as *Couetous* as a Barren-Womb,
the *Graue*, or what's more Ravenous.
B^{l}. K^{t}. We are for you (Sir)
Call you that heynous, that's good husbandrie?
Why, we make money of our *Faithes*: our *Praiers*,
We make the very Death-bed buy her Comforts,
Most deerely pay for all her Pious Councells,
Leave *Rich Reuenewes*, for a few *Weake Orizons*:
or els they passe (vn-reconcilde) without 'em.

Did

1561 *reconcilde*] *d* touched up

65.

Did you but View the Vaultes Within *our Monasteries*
you'ld sweare then *Plutus* (whom the *Fiction* calls
2329 the *Lord of Riches*) were Entombd within 'em.
2331 *wh*. K^t. Is't possible?
2330 B^l. *D*. You cannot Walke for Tuns.
2332 *wh*. *D*. But how shall I bestow the *Vice*, I bring (Sirs?)
You quight forget me: I shalbe shut out
by your strict key of Life.
B^l. K^t. Is yours so vild (Sir?)
wh. *D*. some that are pleasd to make a Wanton on't
Call it *Infirmitie of Blood*, *Flesh Frailetie*,
but (certen) there's a worsse *Name* in your Bookes for't.
B^l. K^t. The Triffle of all Vices: The meere Inocent:
2340 the very Novice of this House of Clay: *Venerie*?
yf I but hug-thee hard, I show the worst on't:
'tis all the *Fruit* we haue here, after Supper.
Nay, at the *Ruines* of a *Nunnerie* once
Six Thousand *Infants* heads; found in a *Fish-Pond*
wh. *D*. how?
B^l. K^t. I how? how Came they thether (thinck you)?
Huldrick, Bishop of Ausberge in's *Epistle*
to *Nicholas the first*, can tell you how.
'may be he was at *Clensing* of *the Pond*.
2350 I can but smile to thinck how it would puzzell

all

1578 *at*] *t* not crossed

all *Mother-Maides*, that ever liv'd in those *Parts*
to know their owne Childes head: But is this all?
B^{l}. *D*. Are you *Ours* yet?
wh. K^{t}. One more; and I am silencd
But this that Comes now will devide vs questionles;
'tis ten times, ten-times worsse then the Fore-runners.
B^{l}. K^{t}. Is it so vild, there is no Name ordain'd for't?
Toads haue their Titles, and Creation gave
Serpents, and Adders those Names to be knowne by.
2360 *wh*. K^{t}. This, (of all others) beares the hiddenst Venom,
2361 the smoothest poyson: I am an *Arch-Dissembler* (Sir)
2364 The Time is yet to Come, that ere I spoake
What my hart meant.
B^{l}. K^{t}. And call you that a Vice?
Avoid all prophanation, I beseech you:
The onely Prime State-Vertue vpon earth.
The *Policie* of *Empires*: oh take heed (Sir)
2370 for feare it take displeasure, and forsake you:
It's like a Iewell of that pretious Value
Whose Worth's not knowne, but to the skillfull Lapidarie:
The *Instrument* that picks-ope *Princes* harts,
and locks vp *Ours* from Them, with the same Motion:
You never came so neere our Soules as now.
2376 ʃ B^{l}. *D*. Now you are a *Brother* to vs.

What

1600 *prophanation*] smeared horizontal bar in loop of *h* 1602 *Empires*] smeared horizontal pen-stroke linked to top left of *E* 1602–5] oblique intermittent pen-strokes, not indicating deletion

B^{l}. K^{t}. What we haue don
hath byn *Dissemblance* ever.
wh. K^{t}. There you lye then:
And the *Game's Ours*; We give Thee *Check-Mate*, by
2380 *Discouerie* (*King*) the Noblest *Mate* of all.
2381 *B^{l}. K.* I am Lost: I am Taken.
\+ *wh. K^{t}. Ambitious,*
\+ *Couetous, Luxurious Falsehood*
\+ *wh. D. Dissembler* includes all.
2382 *B^{l}. K^{t}.* All hope's confounded.
2383 *B^{l}. Q.* miserable Condition:
2386 *wh. King.* Oh, let me bles mine Armes with this Deere *Treasure*,
Truthes glorious Master-Peece; See (*Queene of Sweetnes*)
He's in my Bosom saffe: And this Faire *Structure*
of Comely *Hono*^r^, (his true Blessd *Assistant*.)
2390 *wh. Q.* May their *Integrities* ever posses
that powrefull Sanctuary.
wh. K^{t}. As 'twas a *Game* (Sir)
Won with much hazard, so with much more Triumph,
I gaue him *Check-Mate*, by *Discouerie* (Sir).
wh. King. Obscuritie is now the fittest Fauo^{r}
Falsehood can sue for: It well suites *Perdition*,
It's their best Course, that so haue lost their *Fame*,
to put their heads, into the *Bag* for *Shame*.

And

The Bagg. opens, & the Black-Side put into it

And there behold, the *Bag's* Mouth, like Hell opens
2400 to Take her due: and the *Lost Sons* appeere
greedely gaping for encrease of Fellowship
in *Infamie* (the last Desire of wretches)
Advauncing their *Perdition*-Branded *Fore-heads*
2404 like *Enuies Issue*, or a Bed of Snakes.
2441 So let the *Bag* clease now: (The Fittest womb
for Treacherie, Pride, and *Falsehood*) whilst we (wynner-like)
Destroying (through *Heauens* powre) what would Destroy:
2444 *Welcom our white-knight, with Lowd Peales of Ioy*
\+ *Exeunt.*

2445 **FINIS**

Epilogue, (spoken by the white-Queenes Pawne

My Mistris (the white-Queene) hath sent me forth,
and bad me bowe (thus Lowe) to all of worth.
2450 *That are true frends of the white-house, and Cause,*
which (she hopes) most of this Assembly drawes.
For any els, by Envies-mark, denoted
To those Night-Glow-Wormes (in the Bag) deuoted
where ere they Sit, stand, or in priuat Lurck.
They'll be soone knowne, by their deprauing-worck.

But

69.

But she's assur'd what They'would commit to Bayne,
Her white-Frends Loues, will build-up faire agayne.

SELECTED PAGES FROM BRITISH LIBRARY MANUSCRIPT LANSDOWNE 690

1.

The Induction

Ignatius Loyola: & Erro^{r}.

Ig. ***hah?*** Where? What Angle of the World is this
that I can neither see the *Politique* Face,
nor, with my refinde Nosthrills taste the ffoote-stepps
of any of my *Disciples*? Sons, and heires
as well of my *Designes*, as *Institution*?
I thought they'had spread over the world by this time,
Coverd the Earthes Face, and made darck the Land
like the *Egiptian Grashoppers*.
Heere's too much Light appeeres, shot From the eies
of *Truth*, and *Goodnes* (never yet deflow'rd.)
Sure *They* were neuer *Heere*, Then is their *Monarchie*
vnperfect yet: A iust Reward, I see
for their Ingratitude so long, to Me
(their *Father*, and their *Founder*)
It's not Five yeeres, since I was Saincted by 'em,
Where slept mine *Hono^{r}*, all the time before:
Could they be so forgetfull to Cannonize
their prosperous *Instituto^{r}*? When they'had *Saincted* Me
they found noe Roome in all their *Kallander*

to

16 *Founder)*] top half of bracket missing

2.

to place my *Name*, that should haue remou'd *Princes*,
pull'd the most Eminent *Prelats*, by the Rootes vp
for my deere Com̃ing, to Make way for Me.
'Let every Pettie-*Martir*, and *Sainct-Homilie*
Roch, *Main*, and *Petronell*, (Itch and Ague Curers)
your *Abbesse Aldegund*, and *Cunigund*.
The *widow Marcell*, *Parson Policarpe*,
Sicelie and *Vrsula*, all take place of Me:
And, but for the *Bis-sextile*, or *Leape-yeare*,
(and that's but one in Three) I fall, by chaunce
into the *nine and twentith daie of Februarie*,
there were no Roome els for Me: ' See their Love,
(their Conscience too) to thrust Me (a *Lame Soldier*)
into *Leape-yeare*? My wrath's vp; and (me-thincks)
I could with the first Sillable of my *Name*
blow-vp their *Colledges*. Vp *Error*, Wake:
Father of *Superarrogation*, Rise
It is *Ignatius* calls Thee: (*Loyola*).
Er. what haue you don? oh, I could sleepe in *Ignorance*
im̃ortally, the Slomber is so pleasing.
I saw the brauest Setting for a *Game* now
that ever mine eie fixd on.
Ig. What *Game*, pree-thee?

The

24 *Make*] *e* not fully formed   29 *take*] *e* not fully formed

3.

Er. The Noblest *Game* of all: a ***GAME** **at Chesse***,
betwixt *Our Side*, and the *white-House*: The *Men* set
in their iust *Order*, readie to goe to't.
Ig. Were any of my *Sons* plac'd for the *Game*?
Er. yes; and a *Daughter* too: a *Secular Daughter*,
that plaies the *Black-Queenes Pawne*: *He*, the *Black-Bishop's*.
Ig. yf ever *Powre* could show a *Maystrie* in Thee,
Let it appeere in this.
Er. It's but a *Dreame*,
a Vision, you must thinck.
Ig. I care not what
so I behold the *Children* of my *Cuñing*
and see what Ranck they keepe.

– Enter yᵉ white House & yᵉ Black (as in Order of the Game)

Er. you haue your wish:
Behold, there's the full Nomber of the *Game*,
Kings, and their *Pawnes*, *Queenes*, *Bishop's*, *Knights and Dukes*.
Ig. *Dukes*? They are call'd *Rookes* by some?
Er. Corruptively:
Le Roc the word: Custode de la Roch,
the *Keeper of the Forts*: In whom both *Kings*
repose much Confidence: And for their *Trust*-sake,
Courage, and *worth*, doe well deserve those Titles.
Ig. The Answeare's high: I see my *Son*, & *Daughter*.

Those

56 *Black*] followed by small mark, probably not punctuation

4.

Er. Those are Two *Pawnes*; The *Black-Queenes*, and the *Bishop's*.
Ig. *Pawnes* argue but poore Spiritts, and slight *Preferments*,
not worthie of the name of my *Disciples*.
If I had stood so nigh, I would haue Cut
that *Bishop's* Throate but I'would haue'had his Place,
and told the *Queene* a Loue-Tale in her eare
would make her best Pulsse daunce: There's no Elixer
of Brayne, or Spirit amongst'em.
Er. why? would you haue them play against Themselues?
that's quight against the *Rule of Game* (*Ignatius*)
Ig. pish: I would *Rule* myself: not observe *Rule*.
Er. why then you'would play a *Game* all by yourself.
Ig. I would doe any thing to *Rule* alone:
It's rare to haue the World *Raign'd* in by one.
Er. See 'em anon: and mark 'em in their *Play*.
Observe: (as in a *Daunce*) they glide away.
Ig. Oh, 'with what Longings will this Brest be tost,
Vntill I see this *Great Game*, won, and Lost? – *Exeunt*./

73 *eare*] followed by mark resembling apostrophe

752 B^{l}. Q^{s}. *P*. Come: helpe me with this *Cabinet*, +
and after I haue sindgd theis papers throughly +
I'll tell thee a strange Story. +
B^{l}. K^{ts}. *P*. yf't be sad +
it's well com. +
757–8 B^{l}. Q^{s}. *P*. 'tis not troubled with much Mirth (Sir) – *Exeunt* +

759–60 ***Sce^{a}. 2^{a}. Enter Fat Bishop, & his Pawne*** 507–8

Fat B. *Pawne*. 510
Paw. I attend at Your Great Holynes Service.
Fat B. For Great, I graunt you: But for greately holie,
there the Soile alters: Fat Cathedrall Bodies
have verie often but leane, litle Soules,
much like the *Ladie*, in the *Lobsters* head
a great deale of Shell, and Garbish of all Colours,
But the pure part, that should take wings, and Mount,
is at last gaspe: as if a Man should gape,
770 and from this huge Bulke, let forth a Butterflye 519
Like those big-bellied Mountaines, which the *Poet* +
deliuers, that are brought a-bed with Mowse-Flesh. +
Are my *Bookes* printed (*Pawne*) my last *Inuectiues* 520
' gainst the *Black-House*?
Paw. ready for *Publication*,
for I saw perfect *Bookes*, this Morning (Sir)

fetch

fat B. fetch me a few: which I will instantly
distribute 'mongst the *white-house*. 525
779–80 *Paw*. with all speed (Sir) – *Exit* +
fat B. It's a most lordly life, to raile at ease: 526
Sit, eate, and feed vpon the Fat of one *Kingdome*,
783 and raile vpon an other with the Iuyce on't. 528
790 Of all things I com̃end the *white-house* best +
for plentie and varietie of Victualls: +
When I was one of the *Black-Side* professd, +
my fflesh fell half a Cubitt: 'time to turne +
when mine owne Ribbs revolted: But (to say true) +
I haue no *Preferment* yet, that's suitable 531
to the Greatnes of my *Person*, and my *Parts*.
I graunt I live at ease, For I am made
the *Master* of the *Beds* (the *long acre* of *Beds*)
But there's no *Mari-golds*, that shutts, and opens:
800 *Flowre gentles*: *Venus-Bath*: *Apples* of *Loue*; 536
Pincks, *Hyacinthes*, *Honey-suckles*; *Daffadown-dillies* +
There was a time I had more such *Drabs*, then *Beds*, 537
now I haue more *Beds*, then *Drabs*. 538
Yet there's no eminent *Trader*, deales in hole-Sale, +
but She, and I haue clapt a Bargaine vp +
'let in at *Water-gate*: for which I haue rackd +
my *Tennaunts* Pursse-strings, they haue twangd agen: +
yonder

Yonder *Black-knight*, (the *Fistula* of *Europe*) **– *Enter Bl. Kn*^{t}*.***
whose Disease once, I vndertooke to Cure **& *Bl: B*^{p}*.***
with a high-holborne Halter: When he last
vouchsaffd to peep into my *Priuiledgd Lodgings*,
He saw good store of *Plate* there, and rich *Hangings*,
He knew I brought none to the *white house* with me,
I haue not lost the vse of my *Profession*
since I turnd *white-house Bishop*. **– *Enter Pawne w*th *Bookes:***
B^{l}. *K*^{t}. Looke: more *Bookes* yet:
yond greazie-Turne-Coate, Gurmandizing-*Prelat*,
doth work our *House* more mischeif by his *Scripts*
(his Fat, and fullsom Volumes,)
then the whole Body of the *Aduerse-Partie*
B^{l}. *B*. oh, 'twere a Master-peece of Serpent Subtletie
to fetch him on this *Side* agen
B^{l}. *K*^{t}. and then dampne him
into the *Bag* for ever: or expose him
against the *Aduerse Part* (which now he Feedes vpon)
and that would doble Dampne him: My *Reuenge*
hath prompted me already: I'll confound him
on both *Sides*, for the Phisick he prescribd,
and the base *Surgeon* he provided for Me:
I'll tell you, what a most Vncatholique Iest
he put vpon me once, When my Paine tortur'd me.

He

He told me he had found a present Cure for me,
(which I grew prowd on, and obseru'd him seriously) 560
what thinck you 'twas? Being *Execution* daie,
he showd the *Hang man* to Me, out at Windowe,
(the Common *Hang-man*)
B^{l}. *B*. insufferable.
B^{l}. *K*^{t}. I'll make him the *Baloon*-Ball of the *Churches*,
and both the *Sides* shall tosse him (he lookes like one; 566/-
a Thing swelld-vp with mingled-Drinck, and Vryn, +
and will bownd-well from one *Side* to an other.) +
Come: you shall wryte: Our *Second Bishop* absent, +/56
which hath yet no Imployment in the *Game*, 567
perhaps, nor ever shall: it may be won +
without his Motion: It rests most in *Ours*. +
He shalbe flatterd with *Sede Vacante*: 568
Make him beleeve he comes into his Place,
and that will fetch him, with a veng'ance, to Vs. 570
For I know, Powder is not more Ambitious +
when the Match meetes it, then his Mind, for Mounting, +
as Covetous, and Leacherous. +
--*Enter ye white-House,*
& ye Black-House (seu^{r}ally)
B^{l}. *B*. no more now (Sir 571
both the *Sides* fill. +
wh. *K*. This hath byn lookd for long 572
fat B. the stronger Sting it shootes into the Blood
of the *Black-Aduersarie*: I am ashamd now

I

37.

I was *Theires* ever: What a Lump was I
when I was leadd in Ignorance, and Blindnes?
I must confes I haue all my life-time plaid
the Ffoole till now.
B^{l}. K^{t}. and now he plaies two parts
the *Foole*, and *Knaue*. 580
fat B. There is my *Recantation* in the last leaffe, +
Writ (like a *Ciceronian*) in pure lattyn. +
wh. B. pure honestie: the playner lattin serves then: +
B^{l}. K^{t}. Out on those pestilent *Pamphletts*, those are they +
that wound our *Cause* to the hart. ***– Enter ye white Queens*** +
B^{l}. *B*. here comes more anger. ***Pawne.*** 581
B^{l}. K^{t}. but we come well prouided For this Storme. +
wh. Q. Is this my *Pawne*? She, that should guard our *Person*? 582
or some pale Figure of Deiection,
her Shape vsurping? Sorrow, and Affrightment 584/+
hath prevaild strangely with her. +
wh. Q^{s}. *P. King* of *Integritie*, 585
Queene of the same: And all the *House*, $Professo^{rs}$
of Noble *Candor*, Vncorrupted Iustice,
and Truth of Hart: through my alone *Discouerie*
my life and $Hono^{r}$ wondrously preseru'd,
I bring into your Knowledge, with my Suffrings 590
fearfull Affrightments, and Hart-killing $Terro^{rs}$,

the

588 *through*] *r* deleted with ink cross

1928 *B*^{l}. *K*. Why, *Bishop*: *Knight*, Where's your *Remoues*? yor Trapps?
stand you now idle, in the *heate* of *Game*?
1930 *B*^{l}. *K*^{t}. My life for yours (*Black Soueraigne*) the *Games ours*,
I haue wrought vnder-hand for the *white-Knight*,
and his Brave *Duke*, and find'em Coming both
fat B. then, for their Sanctimonious *Queenes* Surprizall
in that State puzzell, and distracted Hurrey
trust my Arch-Subtletie with.
B^{l}. *K*. oh *Eagle*-pride,
never was *Game* more hopefull of *our Side*. – *Ex*^{t}.
B^{l}. *K*^{t}. yf *Bishop* Bull-beoff, be not Snap'd at next Bowt
1939–40 (as the *Men* stand) I'll neuer trust *Art* more – *Exit*.

1941–7 ***Scc*^{a}. 3^{a}. Enter ye Black Q^{s}. Pawne (wth a Tapo^{r}** *in her hand) and Conducts the* ***white Q^{s}. Pawne*** *(in her Night Attire) into one Chamber: And then Conuaies the* ***Black B^{s}. Pawne*** *(in his Night habit) into an other Chamber: So putts out the Light, and followes him.*
a Domb show

1948 ***Scc*^{a}. 4^{a}. Enter y^{e} white Knight, & white Duke.**

---

wh. K^{t}. True Noble *Duke*, (Faire *Vertues* most endeer'd one,)
1950 let vs prevent their ranck Insinuation
with truth of *Cause*, and *Courage*: meete their Plotts
with Confident *Goodnes*, that shall strike'em groveling.

Sir

1186, 1191 *Scc*^{a}.] sic

\+ of their fixd Vertues, could you let in Seas 1280
of populous Vntruthes against that *Fort*,
'twould burst the prowdest Billowes.
wh. Q. my ffeare's past then: 1283
wh. K. ffeare? You were neuer guilty of an Iniury +
to *Goodnes* but in that. +
wh. Q. it staid not with me (Sir) +
wh. K. It was too much, if it vsurpd a Thought: +
' place a good Guard there. +
wh. Q. Confidence is sett (Sir). +
wh. K. Take that *Prize* hence: Goe (*Reuerend of Men*) 1284
put *Couetousnes* into the *Bag* agen.
fat B. The *Bag* had need be sound, or it goes to wrack,
\+ Sin, and my Waight, will make a strong-one Crack – *Exeunt* 1287–8

2001 Actus Quintus.

2002–3 ***Sce^{a}. pri^{a}. Enter Y^{e} Black-Knight (in his Litto^{r}) & y^{e}*** 1290–2
\+ ***Black-B^{s}. Pawne (above)***

2005 *B^{l}. K^{t}.* **Is** the *Black-Bishop's Pawne*, the *Iesuite*, 1294
planted above, for his Concise *Oration*?
B^{l}. B^{s}. P. Ecce Triumphanti Me fixum Cæsaris Arce.
B^{l}. K^{t}. Ar't there (my holy *Boy*)? Sirha: *Bishop Tumbrell*,

is

1289 *Quintus*] followed by pencil cross above

is Snap'd i'th' *Bag* by this time
2010 *B*^{l}. *B*^{s}. *P. Hæretici pereant sic.*
B^{l}. *K*^{t}. All Lattin? sure the *oration* hath infected him:
2012 Away: make haste: They are Coming.

Enter Black House, meeting y^{e} ***White-K***^{t}***. & Duke***

2017 *B*^{l}. *B*^{s}. *P. Si quid mortalibus vnquam oculis hilarem, et gratum aperuit Diem: Si quid per amantibus Amicorum Animis gaudium attulit, peperituè Lætitiam (Eques Candidissime-prælucentissime) fælicem profectò Tuum, a Domo Candoris, ad Domum Nigritudinis Accessum, promisisse, peperisse, attulisse fatemur. Omnes Aduentus Tui conflagrantissimi: Omni qua possumus Lætitia, Gaudio, Congratulatione, Acclamatione, Animis obseruantissimis, Affectibus diuotissimis, obsequijs Venerabundis Te Sospitem*
2027 *congratulamur.*
B^{l}. *K.* Sir) In this short Congratulatorie Speech
You may Conceive how the whole *House* affects you.
2030 *B*^{l}. *K*^{t}. The *Colledges*, and sanctimonious *Seed-Plotts*
\+ *wh. K*^{t}. 'tis cleere, and so acknowledg'd (Roiall Sir)
B^{l}. *K.* What Hono^{rs}, Pleasures, Rarities, Delightes +
your Noble Thought can thinck. +
B^{l}. *Q.* your faire eie fix on +
that's comprehended in the spatious Circuit +
of our *Black-Kingdome*, they are your Servants all. +
wh. K^{t}. how amply you endeere vs? +
\+ *wh. D.* they are Fauo^{rs} +

that

85.

that equally enrich the Roiall Giver,
as the *Receivo*^{r}, in the free Donation.

B^{l}. *K*^{t}. Harck: (to enlarge your Wellcom) ffrom all Parts
is heard sweet sounding Ayres: Abstruse things open
of voluntary Freenes: and yond Altar
(the Seate of Adoration) seemes t'adore
the Vertues you bring with you.

Musick

*an **Altar** discouered w*th ***Tapers** on it: and diuers **Images** about it*

wh. *K*^{t}. there's a taste
of the old vessell still:

wh. *D*. the *Erronious Rellish*.

: ***Song*** :

Wonder, work some strange delight
(this Place was neuer yet without)
to wellcom the faire-white-house Knight,
and to bring our hopes about.
*Maie from the **Altar Flames** aspire,*
*those **Tapers** sett them selues on fire*
*maie senceles-**Things** our Ioies approue*
*and thos **Brazen-Statues** moue,*
quicken'd by some Powre aboue.
or what more strange, to show our Loue.

***The Images** moue in a Daunce*

B^{l}. *K*^{t}. A happie Omen waytes vpon this Howre,
All *moue* portentously, *the right hand* way.

B^{l}. *K*. Come: Let's sett free all the most Choice Delightes
that ever adornd Daies, or quickend Nightes. – *Exeunt*.

I

THE PART OF 'POORE'

This edition of the part of 'Poore', from a manuscript in the Harvard Theatre Collection, Harvard College Library, was prepared by David Carnegie, assisted and checked by G. Blakemore Evans and John Creaser.

November 1992 N. W. Bawcutt

INTRODUCTION

THE actor's part for the character 'Poore' is apparently the only record of an anonymous and otherwise lost play, here referred to as the *Play of Poore*. The part of Poore is found in Harvard MS. Thr. 10.1, a gift of Mr and Mrs John F. Fleming in 1960. The manuscript book is a sextodecimo in generally good condition measuring 147 mm by 98 mm. It is a collection of four actors' parts from plays written and performed at Christ Church, Oxford, in the early seventeenth century. A fuller description of the manuscript will be found in David Carnegie's 'Actors' Parts and the "Play of Poore"', *Harvard Library Bulletin*, xxx (1982), 5–24, and in Carnegie's 'The Identification of the Hand of Thomas Goffe, Academic Dramatist and Actor', *The Library*, 5th ser. xxvi (1971), 161–5.

The part of Poore occupies fols. 21–46, written on both sides of the leaves, usually about thirty lines to the page. The folio numbers were added, in pencil, by a later hand. This part, like the others in the manuscript, appears to have been copied into an already bound blank book, as the text continues from one gathering to the next uninterrupted, with the only blank pages (between parts, and fols. 44^{b} and 45^{a}) occurring in mid-gathering, and also as the writing is cramped near the binding (e.g. 176, 297, 1095, 1268, 1393, 1578–9). Margins are narrow and irregular on both sides, and speech prefixes often encroach on the text. Act and scene headings are usually supplied for those scenes in which Poore appears. The general form of the part is as follows: a speech prefix for a character other than Poore, followed by a horizontal rule drawn to about half a verse-line's length, followed by the last half verse-line of that character's speech; a speech prefix for Poore, followed by his complete speech; a speech prefix for another character with rule and cue for Poore; and so on. The part is written in a neat italic hand that has not been identified. (At 335 there is a correction in another hand, in a secretary style.) One of the other hands in the manuscript was established by David Carnegie in 'The Identification of the Hand of Thomas Goffe', and it is concluded there that the part of Poore is in a different hand. However, until very detailed investigation of the hands can be undertaken, it may be necessary to leave open for the moment the slender possibility that the hand might be Goffe's.

One ink is used throughout, with a change of pen on the last page. There are only a few uncertainties. The initial letter L has the same majuscule form and the initial letter Y the same minuscule form in both minuscule and majuscule positions, and majuscule K, O, S, V, and W are often only distinguishable from their minuscules by size. Ligatured æ (e.g. Scæna, 1, 187, 410, 429, etc.) and œ (Phœbus, 1307, 1312) are indistinguishable. A tendency to leave space between syllables renders word-division difficult to ascertain on occasion. Several terminal punctuation marks are irregularly formed (e.g. 145, 149, 188, 656, 725). Evidence of the scribe's running out of space at the edge of the page occurs at 3, 10, 348, 469, 513, 515, 723, 798, 984, 1368, 1540–2 and 1577 as well as at the examples of cramped writing near the binding, referred to above. Some of these cases involve turned-over lines. The scribe was careful about indenting Poore's lines if they constituted the completion of a verse line already started by the previous speaker, but sometimes he started writing at the left-hand margin, realized the error, and started again with the necessary indentation. This anticipation is found at 70, 75, 310, 319, 325, 329, 356, 608, 667, 713, 853, 939, 1044, 1240 and 1441. Frequently blots obscure one or more letters: 23, 25, 37, 64, etc. Spelling is sometimes inconsistent, the most noticeable example being *Cytty* (25), *Cyty* (522) and *citty* (527) (for 'city'). Others include *tearmes* and *termes*, *heare* and *heere* (for 'here'), *thanke* and *thancke*, *sicknes* and *sickenes*. Nevertheless, the care and good sense of both copying and alterations suggest an intelligent knowledge of the play; the likelihood that the scribe was either the author or the actor is discussed in Carnegie, 'Actors' Parts and the "Play of Poore"'.

The one passage where serious doubt arises is at 591–4. Despite the speech prefix *Hard* at 591, *Hee's fire and toe, I doe instruct you savingly* is clearly the end of Poore's speech beginning at 585. A horizontal rule has been drawn between 592 and 593 rising to a vertical squiggle after *savingly*. The words *Not aske her portion!* are written small and interlined below *Hee's fire and toe* and above the rule, though the rule does run through *portion!* without appearing intended to delete it. ⟨*P.*⟩ is uncertain; the thick horizontal rule obscures much of the letter. ⟨*P.*⟩ *Yes you may enquire but not &cæ* is written very small to squeeze it into the space remaining to the right of *promiseth*, and *not &cæ* is interlined under *Yes you may*. The passage may probably be reconstructed as follows:

Poore: . . . *or elce*
Hee's fire and toe, I doe instruct you savingly.
Hard: ————————*Not aske her portion!*
Poore: Yes you may enquire but not aske her portion.
Hard: ————————*of what hee promiseth.*
Poore: Hir fathers bond . . .

Some pencil marks of two principal types occur in the left and occasionally right margins. One type is a series of short vertical lines occurring intermittently beside speeches or parts of speeches (82–90, 346–51, 415–20, 421–3, 513–16, 715–35, 899–903, 1094–7, 1102–7, 1123–53, 1306–15, 1412–14, 1450–8, 1470, 1481–2, 1485–7, 1495–6, 1574–8). The other type is a crossed dash beside a line of verse (63, 286, 360–1, 367, 526, 553, 570–1, 574, 597, 598, 729, 774, 775, 811, 819, 845, 898, 958, 959, 960, 1065, 1071, 1098, 1116 [twice], 1239, 1241, 1376, 1381, 1388, 1418, 1527, 1528). It is not clear what the pencil marks signify; some of the vertical strokes are opposite passages of heightened rhetoric, but not consistently so. Furthermore, it is not known whether the pencil markings are contemporary with the manuscript. One word in the manuscript (*not*, 1255) may be in pencil.

That the part of Poore was copied out for (possibly by) the actor for performance seems clear both from its format and from its place in a collection of such parts. The close association of these parts with theatrical performance at Christ Church, and a conjectural reconstruction of the plot of the entire *Play of Poore*, are discussed in Carnegie, 'Actors' Parts and the "Play of Poore"'. The same article suggests a year or two prior to 1617/18 as the most likely date for the part of Poore.

The present edition is the first publication of the part of Poore, and permission of the Harvard Theatre Collection, Harvard College Library, is gratefully acknowledged. The preparation of this edition has also been assisted by research grants from Otago and McGill Universities. Square brackets indicate deletions and angle brackets enclose questionably legible portions of the text. A uniform margin has been adopted at the left-hand side of the page for speakers' names and full lines of text. Act and scene headings have been printed in slightly larger type in order to make them more visible, but there is no distinction of size in the manuscript.

THE PART OF 'POORE'

List of Characters
(in order of appearance)

POORE, an impoverished scholar and trickster (later disguised as 'CHANGE', a Yorkshireman)

SLY, Poore's accomplice (later disguised as GASPAR, a servant, and as Gill's Yorkshire 'UNCLE [CHANGE]')

STRANGE, Poore's uncle, not revealed as such until Act V

GILL, accomplice of Poore and Sly, ? and Sly's sister (later disguised as 'MADAM CHANGE', daughter of a Yorkshire Justice of the Peace)

HARD, a usurer, uncle of Trugull

THIRD OFFICER

QUICKE, enamoured of Snaile's Wife, ? and a lawyer

TRUGULL, Hard's foolish young nephew, enamoured of Gill (as 'MADAM CHANGE')

SNAILE, a cuckold, ? and a merchant

MEDLE, lover of Snaile's Wife, a lawyer, a scholar-hater

SNAILE'S WIFE

DRY, ? confidante, nurse or servant to Snaile's Wife

BADGER, Strange's servant

Two other officers are implied by the presence of Third Officer.

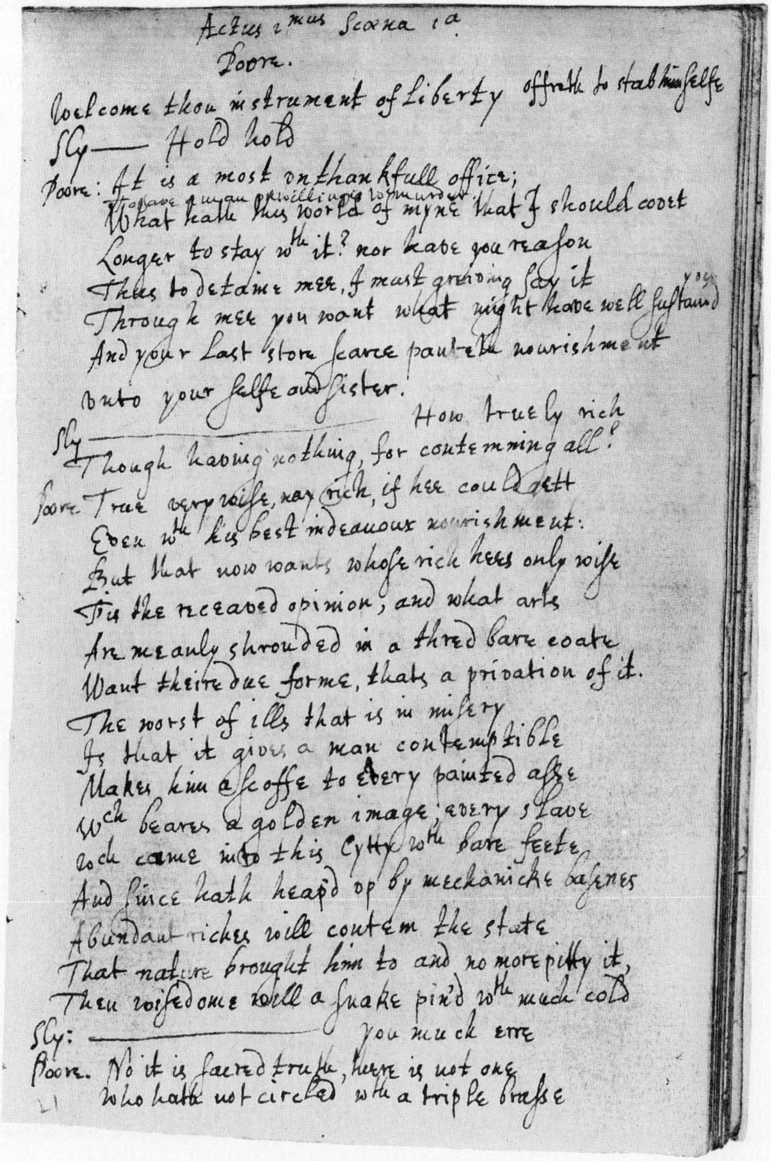

Actus 1mus Scæna 1a.
Poore.
Welcome thou instrument of Liberty offreth to stab himselfe
Sly—— Hold hold
Poore: It is a most unthankfull office;
To save a man unwilling is to murder.
What hath this world of myne that I should court
Longer to stay wth it? nor have you reason
Thus to detaine mee, I must greiving say it
Through mee you want what might have well sustaind you
And your last store scarce paulethe nourishment
unto your selfe and sister.
Sly —— How truely rich
Though having nothing, for contemning all!
Poore. True very wise, nay rich, if hee could gett
Even wth his best indeavour nourishment:
But that now wants whose rich hees only wise
Tis the receaved opinion, and what arts
Are meanly shrouded in a thred bare coate
Want theire due forme, thats a privation of it.
The worst of ills that is in misery
Is that it gives a man contemptible
Makes him a scoffe to every painted asse
Wch beares a golden image, every slave
wch came into this Cytty wth bare feete
And since hath heap'd up by mechanicke basenes
Abundant riches will contemn the state
That nature brought him to and no more pitty it,
Then wisedome will a snake pin'd wth much cold
Sly: —— you much erre
Poore. No it is sacred truth, here is not one
21 who hath not circled wth a triple brasse

THE PART OF 'POORE', FOL. 21a

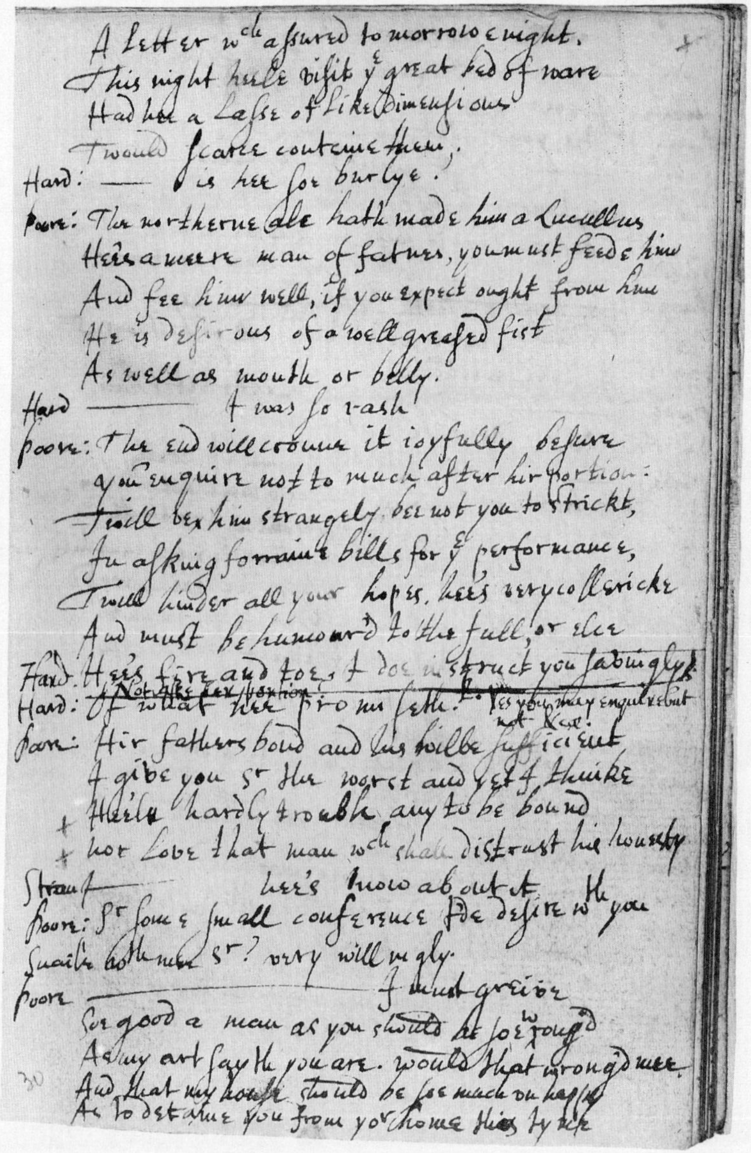

A letter wch assured to morrowe night.
This night heele visit ye great bed of ware
Had hee a Lasse of like dimensions
Twould scarce conteine them.
Hard: —— is hee for burlye.
Poore: The northerne ale hath made him a Lucullus
Hees a meere man of fatnes, you must feede him
And fee him well, if you expect ought from him
He is desirous of a well greased fist
As well as mouth or belly.
Hard —————— I was so rash
Poore: The end will crowne it ioyfully besure
you enquire not to much after hir portion:
Twill vex him strangely bee not you to strickt,
In asking forraine bills for ye performance,
Twill hinder all your hopes, hees very collericke
And must be humourd to the full, or else
Hard. Hees fire and toe, I doe instruct you fadingly
Hard: Of what hee ~~from fathe~~. Yes you may enquire but not ~~Not like her portion~~
Poore: Hir fathers bond and his wilbe sufficient
I give you Sr the worst and yet I thinke
+ Heele hardly trouble any to be bound
+ not Love that man wch shall distrust his honesty
Straunt —————— hee's now about it
Poore: Sr some small conference Ide desire wth you
Snaile both ~~mee~~ Sr? very willingly.
Poore —————————— I must greive
Sor good a man as you should be soe wrongd
As my art sayth you are. would that wrongd mee.
30 And that my house should be soe much unhappy
As to detaine you from yor home this tyme

THE PART OF 'POORE', FOL. 30a

THE PART OF 'POORE'

Actus 1^{mus} Scæna 1^{a}. [Fol. 21a]

Poore.

Welcome thou instrument of liberty   offreth to stab himselfe
Sly   —— Hold hold
Poore: It is a most vnthankfull office;
To save a man vnwilling is to murder.
What hath this world of myne that I should covet
Longer to stay wth it? nor have you reason
Thus to detaine mee, I must greiving say it
Through mee you want what might have well sustaind you
And your last store scarce panteth nourishment
Vnto your selfe and sister.
Sly   ———————— How truely rich
Though having nothing, for contemning all?
Poore.   True very wise, nay rich, if hee could gett
Even wth his best indeauour nourishment:
But that now wants whose rich hees only wise
T'is the receaved opinion, and what arts
Are meanly shrouded in a thred bare coate
Want theire due forme, thats a privation of it.
The worst of ills that is in misery
Is that it gives a man contemptible
Makes him a scoffe to every painted asse
W^{ch} beares a golden image, every slave
W^{ch} came into this Cytty wth bare feete
And since hath heap'd vp by mechanicke basenes
Abundant riches will contem the state
That nature brought him to and no more pitty it,
Then wisedome will a snake pin'd wth much cold
Sly: ———————— you much erre
Poore.   No it is sacred truth, there is not one

3 *liberty*] *l* written as majuscule; see Introduction   s.d. slightly above line and cramped   6 interlined between 5 and 7   10 ²*you*] interlined above *sustaind*   11 *panteth*] *n* formed as *u*, also *plant*, 315   23 *every*] *v* blotted   25 *came*] *a* blotted, perhaps an alteration from *o*

Who hath not circled wth a triple brasse
His more obdurate heart, each man doth live [Fol. 21b]
As hee were enemy to the whole world.
There is a spatious distance twixt the heart,
And tounge of every man, they speake and doe
Nought that hath smallest coherence wth theire minds;
They doe even strive vnto it wth theire full nerves.
Sly ———————— Imitate theire manners.
Poore: You advise well, I shall, and digg a prey
From out theire frosen intrailes, wch shall nourish vs,
Feede vs wth laughter, cramm vs full wth gold.
I'le hold as firme antipathy wth men,
As doe the elements amongst themselves.
Sly ———————— they doe generate
Poore: Soe will not I vnlesse a misery,
And wanton spleene to laugh at it.
Sly ———— will force frequent troops
Of clyents, to your lure. ——
Poore:—— And being well lured,
Ile cramm them soe they shall not breath to flight.
Let's see they may doe well if more harsh fate
Bite not our blooming fortunes.
Strange ———————— beene ith fashion to
Poore. Whilst wee, Apollo's children, wch are given
To the true study of whats purely good,
Share not the least part of it in effect.
Our merits are defects, and only staines,
Disgraces to mans glosse, in mans false eyes.
The heaven of our glory shines no more,
Then a faint candles light, in a proud sunn.
Oh Iove! oh Iove! why hast thou warn'd thy thunder[?]

37 *coherence*] [1]*e* blotted 40 *prey*] *e* altered from *a* 43 *men,*] followed by what may be a point 49 *Of*] *O* might be minuscule 50 rule inserted later to indicate completion of verse-line 49 62 *thunder.*] heavy vertical deletion above point; probably a question mark

It should not dare to touch Apollo's tree?
Yet sufferest vilder more inferiour stro⟨a⟩kes
To rend, and hammer his more loved children, [FOL. 22a]
To dust, to aire, to nothing, lesse then nothing.
Strang: [f] ———————————— for what they suffer
Poore: S^{r} I have fellowe feeling of theire ills.
Strang ———————————— tis sacred truth.————
Poore: [O] ——————— O S^{r} beleeve him not
He doth intice you to a dangerous ill
Sly: Slight what doe you meane? —————
Poore ——————————— Hee is a strange hyæna
Sly: You wont vndoe your selfe —————
Poore [A] ———————— And drawes you on.
Stra: ————— wants much connexion
Poore To losse ——————————
Strange: of what? ——————
Poore:—— Your wealth and reputation.
Riches are not more enimyes to heaven,
then To our art.
Sly ——— honest men in as bare naps.
Poore Our heaven of poetry cannot brooke such rivals,
It is wellnigh[] prodigious they should meete,
And or proceedes from a defect of wo⟨r⟩th,
Or by excesse of some vild humour ioyned,
W^{ch} naturalists observe wthin theire subiects
To cause a vitious forme; for more then perfect
Is but a plurisy wch in wholsomest blood
Breeds naught save malladyes, but being ill,
It meerely is necessited to kill.

64 *stro⟨a⟩kes*] ⟨*a*⟩ blotted, possibly deleted 68 *Poore:*] colon uncertain, obscured by descender of *g* in *Strang* above 73 *Hee*] there are what appear to be apostrophes over both *e*'s 79 rule inserted later to indicate a continued verse line *wealth*] *a* altered from *l* 81 *then To*] *then* is to left of normal margin and *To* capitalized, and was added after line started 82 *in as*] a mark like an inverted *V* above and between these two words 83 *heaven*] *v* blotted 84 *wellnigh*] heavy deletion of a letter, perhaps *t*, added to this word 85 *wo⟨r⟩th*] ? *t* altered to *r* 91 *meerely*] gap between ^{3}e and *l*

You knowe the daunger S^{r} If you proceede
Strange: ——————— You cannot fright mee.
Poore Now comes your cue to speake goe on and roundly
Sly ——————— not shewe his matchlesse skill
Poore: You may proceed and hee may ⟨w⟩inn by intising.
But by your pardon, you are much [deceaved] vnwise, [FOL. 22b]
If all his traines cann lead you to consent.
Strange: ——————— vnto your art
I cannot be disswaded ———————.
Poore ——————— then resolve
To contemplation, for you must neglect
All worldly matters, and be given to this,
As to the sollidst earthly happinesse.
Strang ——————— you knowe my minde
Poore: And I will vndertake to give instructions
In this quainte rhetoricke, and subtile logicke,
And what I cann participat in naturals
Shall not be wanting, since I knowe you firme
Of good capacity and ingenuous.
Strange: ——————— What I possesse
Shall not be wanting to you
Poore: [Pish] ——— pish no no you shall not,
Those were but by words wch I did obiect
Sly: ——— into your minde, I told you soe.
Poore S^{r} It was ill donn, and no way worth your thanks.
Strange: [I would] lodge heare about ———————
Poore ——————— Twilbe best
Strange ——————— only take this as earnest
Poore: It should not neede but since you'l have it soe
I will accept it and deserve it to
Strange: Till when I leave you.———

97 *vnwise*] interlined above deleted *deceaved* 106 *Poore:*] colon uncertain, obscured by descender of *g* in *Strang* above 110 *ingenuous*] ^{2}*n* altered 117 [*I would*]] ? rule through words not intended as deletion but as indication of first half of cue line; cf. deletion at 113 120 *you'l*] *'l* altered from *r*

Poore ——— pray good s^{r} your name
Strange: Tis Strange anon Ile come.———
Poore ——— you shalbe welcome.
Sly —— to quircks and quillets soe they'de help to thrive
Poore: S'light what doe you meane?
Sly ——— my tender Iuvenall
Poore: You wo'n't vndoe your selfe?
Sly — wth your precisenes.
Poore may you have game and will not sterve and perish?
Sly ——————— Leave it scholler leave it [FOL. 23a]
Or it spoile thee
Poore: [] ——You'r spoild you may turne ballad munger.
Sly: Prethee vrge these no more ———
Poore: —— you may thrive, tis possible,
But Ive seene honest men in as bare naps.
Sly ———— Ile pay thee for it
Poore: Doe spare mee not, I will indure thy worst,
And answer thee wth full as great a noyse.
My flash shalbe as violent and as horrid.
Sly: Our lightning shall insue ———
Poore: ———— content content.
Now my wise wench of brantford, how now Gill,
What newes bringst thou now?
Sly: Wee are quite vndon
Poore: On wth your night gowne Gill, and dresse yo^{r}selfe
Ith lady fashion speedily, and returne.
Theire coming in?
Gill ——— I I ———
Poore — Begonn, be gonn.
Sly: —— as poore indeed as thou in name
Poore: Your witt is rich enough to play on mee

126 *and*] *d* altered 129 *wo'n't*] first apostrophe uncertain, open looped head; cf. looped apostrophes of *vpon't* (159) and *you've* (216) 130 *precisenes*] ²*s* appears to be altered from *d* 134 letter, probably *Y*, started preceding rule 139 *Doe*] *D* over-written *thy*] *y* written over *e* 141 *violent*] *o* not closed 142 *insue*] space between *n* and *s* 147 *Gill*] ? *G* altered from *g* 148 *and*] *d* retraced

Sly ——————— I will stab my selfe
Poore That shall not be indited for your death
⟨Ieamy⟩ hath putt it vp [⟨yyo⟩] you shall not have it
Sly: Then Ile goe hang my self:————————.
Poore — Away away man
What what in desperation, fy vpon't
Heare mee s^{r} I have heard a cunning hand
May soe dispose two glasses as by them
Each externe inconvenience maybe kend.
Sly: —— laugh[ ]t at my afflictions? [Fol. 23b]
Poore. At thy promotion, at thy exaltation.
Giv'd thou mayest cheat securely free of feare.
Thou feelst the worst of it, false dice, halfe cards
Will doe exceeding well; [f⟨a⟩lse] if thoult be honest,
Ile teach thee a more exquisite art of begging,
Then ere was heard yet from the newgate dungeon.
Each man ith house shall give a groat a day
To have thee vndergoe theire worke, and gaine by it.
For I will vndertake, in halfe a yeare.
Thou shalt as palpably snatch from the grates,
Of all the prisons wthin London walls,
Ney and the libertyes, the penny pension
As the Kings men doe from theire neighbour companyes
Societyes of gallants
Sly ——————————— death and damnation
Poore Hell and misery ——————————
Sly: —— light on the head,
Poore ———— of thy destroying Hard.
I'st Is't I heare them, fly and putt you on

156 ⟨*Ieamy*⟩] legibility uncertain; possibly *Ieanny* (minim missing, but cf. 250n.) *vp*] followed by deletion of what appears to be a first attempt at *you* 160 *cunning*] *u* altered ? from *o* 163 *laugh*[ ]*t*] vertical stroke before *t* deleted 166 *it,*] comma uncertain, obscured by ascender of *l* in deleted *f*⟨*a*⟩*lse* below 167 [*f*⟨*a*⟩*lse*]] probably scribal eye-skip from 166 above 174 *walls,*] comma uncertain, obscured by ascender of *s* in *pension* below 176 *companyes*] *m* blotted; *es* interlined above *ny* 181 *Hard*] *H* altered, possibly from *h* (cf. 258)

Some other shape, come Lady Gillian come
Have you not donn yett? oh your well enough
Good morrowe to your worships Ladyship-
Good Madam Change.
Act: 1^{I} Scæna 2da
3^{d} offi: some kind purgation, is not that your meaning?
Poore: Madam doe you still hold those points of complement
In wch I did instruct you yesterday?
When to advance, when to retire, and when
To keepe your stand? at the first salutacion
How to congratulate the welcome of—
—A freind equall in fortune, of a superiour,
How to be court⟨e⟩ous to inferiours?
Gill — wth a greater matter [Fol. 24a]
Poore: Thus farr weeve gonn i'th science, theory,
Now weele proceede vnto the art, or practise.
Hard —— —— wee shall see fine sport.
Poore: Thinke you, you cann performe what I instructed?
Gill ———————— make experience S^{r}
Poore Suppose mee, lordly, after what manner meete you
Gill ———— vnworthy roofe of ours.
Poore: How to a knight your equall? ————
Hard — I would my kinsman had hir
Poore: Soe would I to: but for the inferiour now?
Hard ————— should be hir ioynter
Poore: S^{r} you bid fairely for hir, you shall have hir.
Your cousen goose shall have hir
Gill — If I cann helpe it
Poore: ——— come come mind them not.
Soe now you are instructed, let us spend
Some tyme, in matters of a more import.
Madam I knowe your birth, and your deservings,

187 *I^{I}*] for *I^{i}*, representing the genitive ending of *primi* 195 *court⟨e⟩ous*] ⟨*e*⟩ altered from *i*
210 *If*] rule after speech-prefix runs into *I*

But what your fortunes are Ive beene content
Yet not to seeke, but now you've given your selfe
Wholly to mee and doe repose alone
Vpon my choyce, I wilbe bold to inquire
That I may neither loose you on a man
Belowe your selfe in merits or in fortune
Gill. Heaven blesse vs what are you? ———
Poore: ——— Murder, murder
Roalfe Gaspar Thomas where are these varlets trowe?
Sly ——— you shalbe held doe you heare.
Poore: What are you? speake, to what end doe you come? [FOL. 24b]
Hard ——— arrest that Sly
Poore: How Sly saucy groome? first enter my house
Wth more then two, tis a sufficient riot
And god knowes what you would, but that our eye,
Our happily seeing eye prevented you;
Thanks to the supreame power wch made it happy
To that fore sight, what not a varlett stirr?
You are consenting to, wee might be murdred[,]
And you not heare of it; where are your fellows?.
You'are sometymes dubly diligent, and a word
Wthin our kenn will make you fly like winde
Where are your fellowes? ———
Sly ——— troth S^{r}, I doe not knowe
Poore: What men are these? ———
Gill: ——— nor heare of any thing.
Poore: Ney you shall stay, the Iustice shall decide
Whither your act be lawfull, tmay perchance
Conclude at Tyburne or the newgate dungeon
Besides a publique lash from hence to ye tower
From thence to westmonest⟨e⟩r, spight of your freinds

219 *neither loose*] written as *neithe rloose* 224 *heare.*] point uncertain 231 *made*] *e* written over *i* 241 *stay*] *s* altered 242 *tmay*] *t* uncertain, possibly *&* 244 *from*] *r* altered from *o* 245 *westmonest⟨e⟩r*] ⟨*e*⟩ heavily blotted

Hard: S^{r} I intreat your favour.———
Poore — That were pretty,
To be god knowes frighted wellnigh to death,
Then only intreat favour, that's fine recompence.
If thou beest worth a penny Ile have that
And all thy freinds cann make if they will save thee.
Hard. S^{r} in good fayth I meant no harme ———
Poore. ——————— thats better,
That shalbe t[y]ryed, goe Gaspar fetch the Conestable
Sly. Tak't least hee doe repent;
Poore. ——————————— how! forty pound?
That is a sweet amends, but whats your name?
Hard: Tis Hard and please you S^{r} ——————— [FOL. 25a]
Poore ——————— Hard? m^{r} Sly
Hath often named you wthin my hearing
An honest creditor, and for his sake,
If wth [th] your haust this lady be appeased
Your peace is made; what recompence shee will
You must attone hir wth, or this cannot mee.
Some toy will please hir best, shee is a woman
A diamond ring of twenty marke that's all
Oh shee was frighted much, had shee beene married
Tenn Suttons wealths could not have saved yor life
Hard I would bestowe
Poore: — Vm lett mee see the gold,
[Ile] offer it; oh these are [the f] Sly's attachments and his bonds.
Hard: Good S^{r} they are ———————
Poore: ——————— Madam this gentleman
Presents to you by mee his mediate
Twenty faire angells, and doth hope to appease

250 *penny*] ^{2}*n* a minim short 251 *thee*] ^{2}*e* altered 254 *t⟨y⟩ryed*] *r* interlined above deleted *y* 255 *Tak't*] apostophe and *t* run together *repent;*] punctuation uncertain, possibly comma and small blot 258 ^{2}*Hard*] *H* altered from *h* (cf. 181) 262 *haust*] cf. 1135 and note *this*] *is* altered, ? from *at* 265 *please*] ^{1}*e* uncertain, ? altered 270 *mee*] *m* altered from *s*, ? anticipating *see* 271 *oh*] *o* not closed *Sly's attachments*] interlined above caret over deleted *the f* 275 *appease*] *a* blotted

Wth this bright sacrifice, your incensed minde:
To add by glorious coulour of this gold,
A pleasing tinture, to your late pale cheeke.
Hard: Is shee a lady S^{r} ————
Poore ———— not yet, a knight
Is ready now to bed hir, and but stay's
The coming of some freinds vnto the ceremony.
Sly ———————— oh! it takes rarely
Poore Some five dayes hince.————
Hard ———— And is shee well affected?
Poore No yet the importunitye hir freinds have vsed,
Have made hir yeeld.————————
Hard ———— so much into hir ⟨e⟩state [FOL. 25b]
Poore. I have no reason s^{r}.————
Hard What may hir portion be? ————
Poore ———— Hir father S^{r}
Iustice of peace in Yorkeshire, hath alotted
Three thousand pound wch wthin twice three months
After the day of marriage shall bee payed;
Vpon condition, y^{t} shee shall have ioynter,
After his death, three hundreed pound a yeare.
Hir fathers age and weakenes will not suffer hĩ
Present vnto these nuptials but hee sends
His brother to consum̃ate what he please.
Hard. Then he concludes all.
Poore — All —
Hard ———————— in my behalfe
Poore Shee hath refer'd hir selfe to my dispose
And if I like the gentleman and the tearmes
It shall goe hard but Ile prevaile so much
Hard ———————— shalbee assured hir

277 *coulour*] *r* uncertain, perhaps altered 281 *now*] a short curved vertical stroke precedes the word 287 *Have*] *v* altered from *d* 288 *⟨e⟩state*] first letter blotted 292 *Yorkeshire*] *i* altered from *e* 296 *yeare.*] *a* altered from *e*, or possibly vice versa; point obscured by ascender of *h* of *hĩ* below 297 *hĩ*] tittle badly blotted; scribe cramped at binding

Poore: Tis faire the gentle man concludes it
Hard ——————————— yes
Hee shall —— ———
Poor [] — as I like him it takes effect
If I cann ought.
Hard — your care shalbe requited
Poore: It is requited i⟨n⟩th the very act
If it doe prove succesfully and well
Hard ——— in the meane while plant for battery
Poore: S^{r} If hee be as you have spoken him
Hee shall not come vnwelcome
Gill You'r welcome [to] ——
Poore: [] —— to your cost S^{r} ———
Sly Footra for Hard
Poore now my Sly blewcoat thou how likst thou this [Fol. 26a]
Is it not better then ye dolefull ditty
Of Ile goe hang or stab my selfe
Sly: Of more rich witt ———————————
Poore: [Tis in] —— tis inforced soe now
But better arts were better ways to thrift
Gett you a country gentile habit, hir vncle
You must be nowe.
Gill [Wh] — what shall become of mee
Poor Be neat and spruise as what you have cann make
You ⟨h⟩have a woer coming that shall pay fort
You want not my instructions how to answer
Though hee how to oppose, and sett on you
When fate affords no other way to live
to get a living needs must

310 single vertical stroke precedes rule, ? starting to form *A* 313 *i*⟨⟩th] a heavy blot follows *i*; it is possible a *w* has been altered to *i* 318 *welcome*] rule following is used to delete *to* 319 single vertical stroke, the beginning of a *t*, deleted by rule at start of line 322 *then*] faint 327 *gentile*] *en* altered from *l*⟨⟩ 331 ⟨*h*⟩*have*] *h* interlined above heavily blotted ⟨*h*⟩ 333 *how*] ? understand *now* *and*] *a* altered from *s*, ? anticipating *sett* 334 *fate*] *t* written over another letter, possibly *y* 335 *to get a living needs must*] interlined above deletion in a different hand

Our wits [must list indeavour wee may thrive] strive
Exeunt
Actus 2^{di} scæ 2da

Poore. Whose at the dore who is it?
Sly: He y^{t} desires to bee a scholler ——
Poor ———————— Goe Sly,
Admitt, admitt them: I must scoure my witt
I feare tis spoil'd wth rust tis not acute
Sly: What are you [bett] ready for them? ——
Poore Ready ready.
Surely twas in Domitian's tyme he lived.
That Iuvenal, the wonder of all ages
W^{ch} have beene since, should live soe much vnknow̃
Soe much neglected in his owne tyme, as none
Would grace theire storyes wth his sacred name,
Nor praise them selves, wth giving him due fame.
Yet tis enough wee knowe and wonder at thee
That once thou wert and that thy works shal bee
Worthy long admiration.
Sly: Noe noe hee shall not Mr Poore ——. [FOL. 26b]
Poore [] —— whose there?
Oh S^{r} I cry you mercy, and your freind,
Your welcom please you sitt, I was translating
A poet wch is prince of all his sect
Of Satyrists, theire manners should give them
Princes of men, though fewe there bee are soe;
Twas Iuvenall wch if it please you heare
I will recite.
Strang: ———————— Yes very willingly
[Though fewe there bee are soe]

342 *admitt*] possibly followed by comma, obscured by ascenders of *l* and *d* in *spoil'd* below 348 *soe*] *e* uncertain, blotted *vnknow̃*] scribe cramped at edge of page 349 *Soe*] *S* faint 351 *giving*] *v* closed at top 354 *admiration.*] the period is extended into a short line stretching upwards and to the right 356 deletion probably anticipation of *w* for *whose* 359 *sect*] *t* appears to be written over *s* 365 cf. 361

Poore   Tis thus Ile not repeat the Latine text.
Shall I continue silent &, indure,
The loude vexations Codrus doth procure
Wth his rude Theseus ? shall this man reherse
His gouned scæne and this his mournfull v⟨er⟩se ?
Shall giant Telephus consume his day
And long Orestes æviternall play
Whose margent is repleat, whose very backe
Scapes not the rage but beares an asselike packe.
Shall these I say much endlesse still be read
And only I continue as if dead
Vnto these labours ? shall I only feare
To vex mens organs and to force a teare ? &c
I only made experience what I could.
Quicke   ——————— you've made vs knowe you soe.
Poore   The'ire sudden and they beare no more of weight
Then a small tyme would give.
Strang: It is well vrged ————
Poore: ———— and no way worth deniall.
Quicke   And make a Ioviall meale.
Poore   ————— in the meane while
Weele vse a prety schollers exercise
One shall propose a theame, & each compose
A couple of verses on it as they sitt   [FOL. 27a]
And if the first speake last, the rest shall take
Theire cups of wine a peece to acuat them
Sly   ——————— who doth propose ?
Poore: Each in his order shall, doe you propose.
Sly   —— wine doth cheare the heart.
Poore   You observe method in your very sport

366 *Tis*] *T* formed as block capital and heavily inked in (cf. 646)   367 *&*] possibly altered; different abbreviations used at 378 and 388   368 *vexations*] *n* uncertain, could be *u*   *Codrus*] sic, for *Cordus*   370 *scæne*] *e* of *æ* vestigial, resembling an apostrophe   *v⟨er⟩se ?*] [1]*e* apparently written over *i*   372 *play*] slight vertical stroke after this word   374 *Scapes*] vertical descender precedes *c*   375 *still*] [2]*l* altered, ? from *b* anticipating *be read*   376 *only . . . if*] *y* and *f* retraced

S^{r} for the good report you give of wine
Ile wish you quicker poets, and th⟨at⟩ myne.
Sly — What what more yet ————————
Poore: — ———— who ere it bee admitt him
Quicke — goe call them in ————————
Poore: — —— you shall not neede we[a]'re those
they doe intend. S^{rs} wee must intreat you
Into another roome, there you shall see
What passeth; if't please you disclose yor minde
I will performe what my weake skill can[n] doe
Sly — ——— Ile lead the way
Poore — Ney pray S^{r} goe, wee schollers love no complement
Though wee cann vse it: he hath beene yor guide
And you must followe

Scæna 3^{ia} Enter Poore above

Poor: — A swagerour doe you say one y^{t} hates schollers?
Hee's none of your stage railours on thẽ is hee?
Quicke — ———— Inns of court man that cann raile
Poore: — I would he were a poet, one that daubd
Papers wth greasy lines, wch fall away
From his hoggs head, as sweat doth frõ his body.
Both being excrements, of art, and nature.
Such I doe knowe there are, & would faine meet wth
Ide make thẽ knowe theire mungrill nature could not
Produce a word, lesse vicious then themselves,
And if not borrowed from ye sacred springs. [FOL. 27b]
But tis to matter; Ile give them leave to envy
What is beyond theire reach. but for yor creature,
If I not bafle him in his proper humour
Ile burne my bookes, and turne a lawyers clearke.
But they are neere the doare you shall have sport.

397 *th⟨at⟩*] ⟨*at*⟩ probably altered from *is* 404 *yor*] *^{r}* blotted 407 *schollers*] first attempt at ascender of *h* is above *c* (cf. 411, 524, 1032) 413 *Inns*] cramped, *s* probably inserted later 416 *frõ*] scribe rightly anticipated running out of space 418 *meet*] *m* blotted 422 ^{1}*to*] sic, ? read *no* *envy*] *n* blotted

I must begonn Exit
[Sly]Quicke — worth cherishing ————
Scæna 4ta Enter Poore
Trugull ———— what, is't a hall?
Poore The best our poore house hath. ——
Tru —— Pray whats your name? ——
Hard —— the gentlewoman minded
Poore: Doubt not but you shall well, I like the man
[That]He is a proper man[] y^{t} will tempt much
Besides grave, generous as it seems to mee
Repleat wth worthy qualityes, & though rawe
In Cupids ceremonyes, I must thinke
A few instructions, will give him singular.
Hard Doe you thinke soe S^{r}?
Poor. Yes vndoubtedly
I know hee's very apt:– to bee a gull.
Snaile: — Pray S^{r} lets see the gentlewoman
Poore: You shall Gaspar lead vp these gentlemen
Vnto your m^{rs}
Sly ———— I will ————
Poore: Stay you wth mee S^{r}
Doe you inquire hir minde and bring hir downe
The whilst wee wilbe busy Gaspar lead them.
Trugull: Must not I goe to must I not?
Poore: Not yet
Trugull. ———— and kisse and talke wth hir.
Poore: S^{r} it is best first to have mediates [FOL. 28a]
Shee shall bee brought downe to you
Strange ———— speake lower
Poor: Pray S^{r} may I inquire your name and country?
[Tru: ———— of the name Ime sure]

428 *Quicke*] interlined above deleted *Sly* 432 *Tru*] blot, probably deleting a *P*, at start of cross-stroke of *T* *Pray*] interlined above rule 435 *He*] interlined above deleted *That* *man*] followed by single deleted letter, possibly *s* 436 *generous*] *ne* interlined above deletion and caret 448 *minde*] entire word appears to have been retraced 452 *Trugull*] ²*u* uncertain

Quicke How say you goodman dawe?
Poore Tis a faire living S^{r}
Tru: — But a faire living?
Poore A very rich one S^{r}.
Trug: ——— I cry you mercy
Poore But S^{r} after what fashion would you woe?
Tru: Why are there divers fashions
Poore [Very many.] Yes as in other things
Soe wee're fantasticall in that, ney more.
Your woer is or rampant or couchant:
Your rampant woer, is an angry fellowe
That beares downe all before him should y^{u} heare him,
You'de thinke hee were a souldier by his wounds.
Heele sweare a woman in to love wth him,
Or spend whole vollyes of his oaths in vaine.
Though that doe seldome happen; for his thunder
Battars theire fortresses vntill they fall
Flatt downe before him.
Trug: ——— Is it possible?
Poore S^{r} very true, your couchant, is a creature
W^{ch} sighs and sobs out Hero & Leander,
Or some more mournfull elegyes; and hee goes
Alwayes crosse armed, to shewe his passions.
Tru: ——— I wilbe that woer [FOL. 28b]
Poore: Soe S^{r} but Ile instruct you soe effectually
You shan't neede halfe y^{t} passion. Let mee see
You have a very perfect spericke eye
True Yes Ide be sory elce ———
Poore: ——— And of congruous health
Tru: Yes I am very health full.
Poore ——————— S^{r} the better

464 *divers*] *s* altered from *d* 469 *heare him*] scribe was already saving space and interlined *him* above *heare* 474 *fortresses*] ^{2}*es* altered, *e* uncertain, *s* altered from *e* 479 *goes*] *e* blotted, uncertain 482 *Poore*] ^{2}*o* blotted 484 *spericke*] ? read *sphericke* (i.e. *spheric*)

Your organs are more fitt; for I must teach you
To fix your eye wth iudgement, on an obiect;
And Ile give such a power vnto ye radiature
Emitted from it y^{t} it shall strike hir
More conqueringly then Cupids golden shafte.
At the first sight you shall not speake to hir
But heare are lines wch when shee ginns approach
Ile desire you to reade, & you shall read thẽ.
Say often say you writt them in hir prayse.

Trug: And they are none of myne ———

Poore ——— oh S^{r} the better
You Imitate the gentile fashion
They for the most part only live on others
By borrowing of others, and shall you
As well proportioned for a genltreman,
As amongst them the best, not keep ye fashion?

Quicke ——— will raile on the whole world

Poore: How! feare to ly? then feare to live, all creatures
Doe live by lying

Tru: som live by standing ———

Poore: ——— Indeede I am deceaved,
For some doe live by standing, yet they ly to.

Tru: It may bee soe ———

Poore: And to beginn wth gallants, for nobility [FOL. 29a]
I durst not touch though they should spend themselves
On waxen Images;
Nor cleargy men though they should ly wth scripture,
And vitiate [th] it to adulte[rate]ry.
Have at your gallants, should they pay theire debts
As they doe promise, I knowe some now flants
In cloath of tyshue, y^{t} would be as bare,

489 *teach*] *c* altered, perhaps from *h*   503 *genltreman*] sic   508 *som*] *s* retraced in minuscule form, perhaps over ill-shaped majuscule   513 *durst*] *r* altered   *themselves*] *ves* interlined above *sel*   515 *should*] *u* interlined above *o*   518 *flants*] *n* formed as *u*

As when they first sett foote vpon this land.
These live by falsifying of theire dayes;
Others by mating wth ye Cyty wives
Schollers and lawyers doe' live by theire toungs
And the best ground of schollers sophistry
W^{ch} you may call lyes; but your lawyers toungs
Are strumpets ly wth all men yet they live by them.
Your citty lying is so truly knowne.
As I will not repeate it.
Stran: ———— wth out cessation
Poore: But to goe forward, shee hearing hir praise read
Cann't choose but speake to you, out of hir words
Then must you take occasion, and proce⟨a⟩de.
If I had tyme Ide give you actions
W^{ch} should prove charmes, and drawe hir by ye eares,
Despight all propased antydotes of deafnes.
Tru ———— and speake soe?
Poore You shall most potently, yor eyes shall [sparkle] spread
Such flames of love, as shee shall feare to stirr
Least shee be scorched wth them, yor lips shall move.
Such sphærelike harmony as you shall ravish hir.
Tru: ———— for ravishing [FOL. 29b]
Poore: No, thinke not Ile vrge ought shalbe distastfull
Tru Nay nay you shant deny it. ————
Poore: ———— ⟨Come⟩good S^{r}
Youle wrong mee much, for I have not deservd it.
Quicke ———— and it shalbe kept.
Poore: But S^{r} I must confesse Ive laboured
And donn you more good wth ye gentlewoman.
Then cann this tenn tymes doubled procure mee.

522 *Cyty*] *C* written over *c* 523 *doe'*] sic 527 *citty*] *tt* written over *c* 531 *hir*] ? *r* altered from *s* 532 *proce⟨a⟩de*] *a* uncertain, not in scribe's usual form 534 *y^{e}*] scribe saving space towards edge of paper 535 *propased*] read *proposed* 537 *spread*] interlined above deleted *sparkle* 544 ⟨*Come*⟩] altered

Yet since you offer it soe vnrequested
I doe accept it as sufficient recompence.
For all my labour, not because tis worth them, —
I like your will, farr better then the gift.
Be mindfull that you wrap a ring ith verses.
Tru: Oh I meant that, will not this serve?
Poor ———— it will
Strange: Not very well ————————
Poore ————— be ready they are coming,
S^{r} shall I heare them.
Tru: Attend for these are they. Poore: S^{r} I doe heare.
Tru: That's for the ring ——————
Poore: —————— S^{r} these are very good
Tru: I would shee heard mee
Poore: Doe you vse this often?
Trug: I would shee'de heard them read. ——
Poore ————— S^{r} ift please you,
I will present them to hir.
Gill Greater perfection to them.
Poore —— tickle hir wth prayse.
Tell hir theire good because theire end is good
W^{ch} is to prayse hir.
Hard When comes hir vncle S^{r}?
Poore ———— I did receave
A letter wch assured to morrowe night. [Fol. 30a]
This night heele visit ye great bed of ware
Had hee a lasse of like dimensions
Twould scarce conteine them.
Hard: —— is hee soe burlye?
Poore: The northerne ale hath made him a Lucullus
Hee's a meere man of fatnes, you must feede him
And fee him well, if you expect ought from him
He is desirous of a well greased fist

558 *are*] interlined above caret   560 *Poore:*] the only occasion of a line of Poore's starting in mid-page

As well as mouth or belly.
Hard ——— I was so rash
Poore: The end will croune it ioyfully besure
You'enquire not to much after hir portion:
Twill vex him strangely, bee not you to strickt,
In asking forraine bills for $\overset{e}{y}$ performance,
Twill hinder all your hopes, hee's very collericke
And must be humour'd to the full, or elce
Hard: Hee's fire and toe, I doe instruct you savingly.
Not aske her portion!
Hard: Of what hee promiseth.⟨P.⟩ Yes you may enquire but not &cæ.
Poore: Hir fathers bond and his wilbe sufficient
I give you S^{r} the worst and yet I thinke
Hee'l[e] hardly trouble any to be bound
Nor love that man wch shall distrust his honesty
Stran [I] ——— hee's []now about it
Poore: S^{r} some small conference I'de desire wth you
Snaile Wth mee S^{r}? very willingly.
Poore ——— I must greive
Soe good a man as you should be soe wrong'd
As my art sayth you are. Would that wrong'd mee.
And that my house should be soe much vnhappy
As to detaine you from yor home th⟨i⟩s tyme
Snaile. I have lost nothing have I s^{r}? [FOL. 30b]
Poore: [] ——— A rare iewell
S⟨na⟩ile I ever had ———
Poore: S^{r} tis your wife I meane.
Snaile: ——— Not gonn S^{r} is shee?
Poore: Hir honour hath left hir, for shee hath left

586 *You'enquire*] apostrophe uncertain, formed as ligature 591–4 confused; see Introduction 595 *wilbe*] *w* altered from long *s* 599 *now*] thick vertical stroke precedes this word 603 *wrong'd*] *w* interlined above caret; *g* altered 606 *th⟨i⟩s*] heavy blot follows *h*, perhaps an alteration, but dot clearly visible above 608 single vertical stroke follows speech prefix, probably mistaken beginning of *A* 609 *S⟨na⟩ile*] *n* written over another letter

To bee an honest wife, you knowe on Medle?
Snaile: ———————————— my good cu[]stomer.
Poore: [Hir honour hath left hir for shee]
T should seeme soe he hath go[od]tt yor best ware S^{r}
Snaile: I nere wrongd you ————
Poore ———— nor ere mistrusted him?
Snaile: No on my life. ——
Poore: ——————— nor wife, I knowe it well
Sir hye you home; if you now meet not wth him,
Ile give you such instructions as you shall
In ye named place at further tyme, meanwhile
I knowe a gentleman whom he hath wrongd
Will give his best indeavour, to finde out
The tyme, & to prevent him if you please.
S^{r} I will send the gentleman to morrowe.
Strange: ——————— to what you please
Poore S^{r} I will send the gentleman to morrowe
That shall intrap him.
Snaile ——————— indeede shee told mee soe
Poore: Pray S^{r} be patient heare. ——
Snaile: I pray you S^{r} remember mee ————
Poore Be sure I will; and send the gentleman to morrow morne
By [that] eight o'th []clocke.
Snaile: ——————— heele deale honestly?
Poore If you mistrust him, one you shall thinke faythfull
Choose to this office, I but offer S^{r},
Tis in your will to'accept
Snaile Be not to credulous I did thinke ————
Poore. ———————————— fy fy [FOL. 31a]

614 *cu[]stomer*] *c* perhaps altered; a single letter following *u* has been deleted with a blot 615 repeats 612 616 *go⟨od⟩tt*] *gott* superimposed on *good*, apparently written through eye-slip from *good* directly above in 614 621 *hye*] ? *h* altered 629 *send*] *d* altered from *t* 634 *tomorrow morne*] written slightly smaller and slightly above line; ? added later 635 *o'th*] *t* altered from *f*; or possibly both are intended []*clocke*] deleted letter perhaps *s* 636 *heele*] ^{3}*e* altered from *d*, ? anticipating *deale* 639 *will*] *w* uncertain, second half blotted

blaze not your owne discredite, tis to much
You know't your selfe.
Snaile: ———— but are you sure tis true
Poore I would I were not ————
Hard Tomorrowe night he comes.
Poore ———— yes yes tomorrowe
Tru: ———— wee shalbe married.
Poore I doubt not but you shall
Hard —— you sha'nt soe suddenly
Poore Are you not yet adultus?
Tru: —— what doe you meane
Poore: not yet of age?
Trug: ——— yes that I hope I am
Poore Will you then suffer S^{r} such contradiction?
Lett them determine of you appoint tymes?
Trug: Nay and I will to ————
Poore: —Oh S^{r} been't to feirce
He is your vncle, you doe owe some duty
Or at the least respect ————
Hard ——— A second father to him.
Poore: You must be rul⟨e⟩d, but not to much oreruld
Tru: ———— Ile warrant you
Poore Sr Heele bee gonn ere this be not to violent
Vpon your wife inquire out secretly.
Hard bee his continuall rendez vouz ———
Poore [A]—— and reason.
Gill I must continue Mrs Change ————
Poore. ———— They heare
You must, a iustice of peaces daughter,
Ith north at least
Quicke ———— did you feare us

646 *Tomorrowe*] *T* formed as block capital (cf. 366) 650 *suddenly*] *n* altered 656 *tymes?*] vertical stroke runs through query (cf. 725) 659 mark at end of line, possibly a dash 662 *rul⟨e⟩d*] *e* uncertain, possibly apostrophe 666 *bee*] written over dash 667 *reason.*] point uncertain, obscured by ascender of *h* in *Change* below 671 *least*] *s* retraced, perhaps altered

Poore — Not as Snaile feares meddle, to morrowe morne
You must to him, hee will initiate you
Him selfe in to acquaintance wth his wife
If you shall neede my counsell, Ile instruct you [Fol. 31b]
How to behave your selfe in information
Quicke — to much I feare ———
Poore — ——— no hee must bee inraged
You must add to his fury and augment it
Quicke — Vpon y^{e} least distastfull word ———
Poore — ——— and lett him
Nay if hee be an angry boy weele deale w$^{th'}$ him
And fright him from his roaring humours, wee
Cann talke, bristle, and vaunt, as well as hee.
Exeunt

Actus 3^{ij} scæna 2da

Poore — What cheaters did hee say?
Sly — —— that was the word
Poore — And couldst thou suffer it goe thou'rt a gull
& that huge bulke of thyne those giant limbs
Conteine not any sparke of man wthin them.
Sdeath had I heard him he should have found I had
A thunder in my hand Iove in my voyce
Sly — ——— and sayth vs cheaters
Poore: — Pish tis a puny one easy to performe.
Ile have a duble or a ⟨no⟩ revenge
Vppon my life I think⟨e⟩[t] thou wouldst confess
Vs cheaters should a man inquire of thee.
Sly — Wee are noe better ———
Poore — ——— I thought this, thou lyest
What ere of cheating's in mee it is thyne:
Thou didst intice, coniure mee by our wants

677 *How*] *w* blotted 683 *Nay*] ? *a* altered from *e* 694 *A*] altered from *T* *my*] altered from *his* 696 *puny*] *n* uncertain, blotted 697 ⟨*no*⟩] legibility uncertain 698 *think*⟨*e*⟩[*t*]] ^{2}*t* crossed out

Didst force me too't when I god knowes was minded
Never to suffer more in this vild world.
Sly — But how much in ye insuing. ————
Poore — Doe not vex mee
By all good things I vowe, and will performe it
If ere I learne, y^{t} a like worde be spoken
Thou hearing, suffering it, I will abiure thee; [FOL. 32a]
Leave thee vnto thy selfe & spoile thy hopes
Sly — You may doe as you please —————
Poore — [G] ——— goe to Virginia
To the Bromoodoes, or elce hire my selfe
Vnto the Northwest passage; if these faile:
Turne Poet stageplayer or any thing.
rather then live wth thee, Ile sell my selfe
Vnto a Iewe or worse, an english vserour
Whom have I cheated? only Ive sold Hard
Fishd my young gallant Trugull vexed Snaile
Intic'd my Strange to poetrie, thats poverty:
W^{ch} hee shall surely feele prevented Medle
Drawne blood from Quicke, or at the least will draw it
What act mongst these deserves ye name of cheating
Ist not to gett from vserours charitable?
And to lett him bee wise, y^{t} is not cousned
Whome nature made a foole is against nature
To lett men knowe when others doe them wrong
Is a great Iustice, and worth recompence.
And to make him a poet that would bee one,
Is att the most but to fullfill his vowes.
What to prevent a lawyer since theire knowne
To circumvent all others, but meere equity?
And to take vengeance on who doe defame vs,

705 *world.*] stop has extended downstroke, possibly accidental 714 *Bromoodoes*] understand *Bermudas* 723 *draw it*] interlined above *draw* at edge of page 724 *these*] *s* altered from *d*, ? anticipating *deserves* 725 *charitable?*] vertical stroke through query (cf. 656) 727 *against nature*] written *again stnature*

Soe it bee noble, is allowed to vs
by Martiall lawe, whome have I cheated now
Whom have I cheated now, or against whom
Have I intended more, then may bee donn?
Sly ——— their end maks actions good [FOL. 32b]
Poore. Tis true my Sly. ⟨I'm⟩ in apparrell well,
Sufficient for a petty gentleman
Where is thy rapier?
Sly. What do'est thou intend? ———
Poore — What cannst thou guesse?
Sly Not well ———
Poore — then aske not, for thou shalt not knowe.
Wher ist ———
Sly — above —
Poore If Quicke doe chance come hither,
Stay him till my returne wch shalbe suddaine.
If heele not stay will him, not goe to Snailes
Till I may speake wth him, Gill bring down ye rapier
If Trugull come lett Gill and hee be private,
If hee be earnest, lett him presse hir his.
Gill ——— spirit on his bankes.
Poore Take heede my Dousabell vnto your docke
Looke not to my affaires; take heede yor Trugull
Bee not to hard for you hees a lusty knave
Cann pitch his barr well, shoote his shaft arright
And pay you home my Gill; hee cann ifayth.
Gill That shalbe tryed ———
Poore ——— bee wary and doe well
Prepare yor selfe vnto yor part anon Exit.

---

740 ⟨*I'm*⟩] legibility uncertain, perhaps altered *apparrell*] *ar* altered, ? from *el* 741 *Sufficient*] *S* altered from *f* 753 *private*] *r* written over another letter; two dots above *i* 756 *Dousabell*] *D* written over *d* 759 *pitch*] *t* inserted between *i* and *c* 763 s.d. slightly lower than line, rule runs up to it

THE PART OF 'POORE'

Actus 3^{ij} scæna 3^{ia}

Med — wish hee had not inquired. Enter Poore disguisd
Poore: Oh Mr Medle I have sought you S^{r}
In all your places of retreat.
Me[l]d —— [—] Vnto what end S^{r}
Poore Wee are private heare
Now I will give it you, you knowe one Quicke
An envious raskall one that laboureth
That seeketh causes to defame all men
And if they want his wil's sufficient [FOL. 33a]
For hee defames them; and vniustly iust
Beginns wth his owne intimates; this vild wretch
Hath quite supplanted all yor hopes at Snailes
Med: —— may bee supplanted
Poore: Nay lett it not seeme strange, I know yor hopes
Your more then hopes your much assurance there
Of his wives love, know all occurrances.
And come to tell you y^{t} you are abused
By this same Quicke, who hath, I knowe not how,
But sure it was by some sinister meanes
Found first you lov'd & after whom yo^{u} loved.
Who hath (to what intent I doe not knowe)
Yet sure hee did intend to wrong you by it
Reveal'd the privacy of your love vnto
Hir husband who now truly iealous
Hath giv'n in charge to one of's trusty freinds,
That if you chance to come thither hee should
Much circumspectly watch your haviour
The manner of your language to his wife
And farther y^{t} hee should bee certified
Of your approach wch how suspiciously

768 *Vnto*] partly written over rule inserted in error 773 *they*] *e* altered 776 *hopes*] blot between *e* and *s* 778 *strange*] *g* altered 786 *wrong*] *w* interlined above and slightly before *r* 791 *haviour*] *vi* blotted

Heed take, the very premisies demonstrate.
Your perill may bee much too, hee is desperate,
And I doe thinke will hardly brooke to see you
Wthout much fury, wch though you esteeme not;
Yet poore gentlewoman. ————
Med Advise mee for ye best s^{r} ——
Poore —— trust mee I will
First be reveng'd on Quicke, & if you cann
Make him confess that only enviously
He scandald you for some small wrong you did him. [FOL. 33b]
Then you devise some other means besides
How to confirme hir honesty
Med: your name I pray s^{r}
Poore —— change a Yorkeshire man
Med. S^{r} I am much indebted to you[r lov]e
Poore ———— and I will study [asside
How you shall pay – oh S^{r} humanity
Commaunds this office
Med: Stronglier knitt betweene vs
Poore S^{r} I desire it may, wch to continue
Ile give you intelligence, for I am ye man
Snaile hath appointed as hir overseer
Med I thanke you.
Poore: When you would speake wth mee send to Poor's house
The scholler, I shall heare of it, the tyme
Will not afford mee farther leisure now
S^{r} fare yow well. Exit

Actus 3^{ij} scæna 4ta

Wife — occasion to vnsluce them —— Enter Poore.
Snaile: ———— to whom should I give credite?
Poore To them y^{t} you thinke best deserve it S^{r},

798 *w^{ch}*] *w* altered, perhaps from *t*   *not;*] very cramped, blotted, at edge of page   800 *mee*] *ee* blotted   809 *you[r lov]e*] deletion firmly includes *r* of *your*, presumably should include *e* of *love*; read *you*   810 s.d. underlined, and preceded by sweeping curved bracket   817 *you.*] point uncertain, obscured by ascender of *d* in *would* below   821 s.d. half a line lower

What place commaundss hee in your credulous heart,
That hee should force beleefe against your wife
Shee may be chaster then the mourning aire
Purg'd by the sunn of vitiating mists.
But yet there is a shrewd suspition
Much frequent in your freinds, they think not soe
Ile vowe, I've heard him say y^{t} he hath knowne hir,
But yet how [vn]truly 'tis vnknowne.
Wife — My duty to you.
Poore: — —— your knowledge I desire
S^{r} I doe greive, I chose soe sad a tyme
For the beginning of acquaintance, but [FOL. 34a]
I hope it shall continue wth more ioy.
This is your fault S^{r}, you are to vnkind,
Vnto soe sweete a wife. ——————
Snail — Be very long ——————
Poore: — —— Sr Ile performe it zealously.
I would be private wth you M^{rs} ——
Wife — Bee privat wth mee
Poore — —— I have strong occasions.
Dry: — —— wth hir privatest counsell
Poore: — Then I dare like wise, you knowe Medle?
Wife — True
Poore — And he hath blabd it
Wife — —— as you meane
Poore: — Oh to to truly
Wife — What⟨,⟩ durst ye villaine say soe? ——
Po — [P] —— Positively.
Wife — And soe Ime knowne.
Poore — By him, for hee perceaving
You now begann neglect him, likewise knowing

826 *commaundss*] *ss* uncertain, ? retraced *hee*] *he* blotted *credulous*] *s* uncertain, blotted *heart,*] squiggle above comma, possibly intended as query 830 *shrewd*] blot on tail-end of *d* 832 *vowe,*] blot over comma extending to *I* following 833 *But*] *B* altered, ? from *A* *yet*] interlined above caret 854 *Wife*] *W* altered, ? from *P*

Your love wa[ll]s fully fixed on Quicke, did thinke
No better way to secure you his owne
Then by revealing your intended love
W^{ch} hee hath fully donn; the other to
Not knowing freelyer to settle you
In your newe love, then by displacing Medle;
Hath striven wth great effect to y^{t} performance
Thus have they laboured to supplant each other
Wife But only I have be⟨e⟩ne tript vp ———
Poore ——— most true
Whilst they reioyce in theire high enterprise
And thinke theire wits much good ————
Wife Ile be revenged —— [FOL. 34b]
Poore —You must that Ile performe
I thinke I have allready ——
Dry Vpon my life ————
Poore ——— You shall not finde mee otherwise.
Wife Your love shalbee rewarded ——
Poore —— wth your I hope
That is my only ayme
Dry ———— deserve to have it
Poore And I will keepe it warely, by this
Your envious lovers may bleed each by other
Wife ——— I shall reioyce
Poore Tis like they will ———
Dry ——— no matter lett them sinke
Poore If not Ile soe provide your honour shall
No whitt be impeached
Wife Then I shalbee vnspotted ———
Poore ——— not knowne otherwise
Wife —— be holding to you S^{r}

857 *wa[ll]s*] *s* interlined above deleted ^{2}l 860 *donn*] ^{2}n imperfectly formed, possibly *e* 865 *be⟨e⟩ne*] ⟨*e*⟩ heavily blotted 875 *your*] ? read *yours* 878 *I*] altered 881 *Tis*] *s* altered 884 *be*] *e* uncertain, blotted

Poore Now shall my ignoramus and young witt
Knowe they have found a scholler y^{t} can iearke yẽ.
Who have wee heare my gull & Gillian
What intend they trowe?
Tru: And you S^{r} ———
Poore ——— I returne your complement
Wth ye like wish to you, & y^{t} faire gentlewomã
Wife ——— Ile give my indeavour
Poore And doe not you vse to carreine your selfe?
What fucus have you daubd your face wth, ha?
Thinke you Ile have you vse theise plasterings
And outgoe snakes in monthly casting skinns
Tru Theide looke like eels for all ye world. ——
Poore —Spraule soe
And be more slipery as they are. but s^{r}, [FOL. 35a]
I hope you not intend hir for your wife
Tru: Beleiv't S^{r} but I doe ———
Poore —— beleive't you must not
Tru: ——— Ile aske hir
Poore: You shall not need, for I cann certify you,
I have reserved hir for my selfe.
Tru: —————— be cousned of my wife?
Poore How S^{rrah} cousned, such []an other word
And Ile lopp of a limbe send you to'the' spittle
There to condole your losse. S^{rrah} if your eares
The want of them I mean cann move you ought
Let mee not heare another word but give hir mee.
Tru: [S^{r} I doe love my eares and feare my eares]
It were a prety toy to gett hir from mee

889 *yẽ*] i.e. *them* 890 *gull*] *g* altered from majuscule, anticipating *Gillian* 894 *gentlewomã*] tittle above space following *a* 898 *have*] *v* blotted 900 *all*] deletion under ^{1}l 904 *Beleiv't*] *v* altered 910 *an*] deleted descender precedes *a* 915–16 the scribe evidently copied the first line of Trugull's speech (915) before realizing his mistake and copying the cue line (916). A sweeping curved bracket following the speech prefix links the two lines.

Poor — Are my words toye
Tru: — Ile try what you cann doe
Poore — Marry and shall trips him vp.
Soe s^r you see now in what plight you are
Tru — ———— doe not hurt mee
Poore — On the conditions y^t I shall propose
You are your owne man shee likewise your wife
You shall give mee to hundred pounds to right
My wrongs. ———
Tru — — but trust mee s^r y^{ts} somewhat hard
Poore: — Doe not deny'it for if you doe by this.
Not forty kicks, not 20 luggs by the'are [by his hand he swears]
As many tweaks by the nose, your fower foreteeth
A little finger shall not save your life
At least a maine limbe.
Wife — For my sake a lesse ransom.
Poore — Your commaund.
I must obay, it shalbe but a hundred. [Fol. 35b]
And heare you [brin] leave it wth yor tutor Poore
Be sure you faile not, if you doe you knowe.
Tru: — [W]— when shall I carry it?
Poore — This night I knowe y^u cann whẽ it please you.
Tru: — I will S^r ——
Poore — —— Gill how goe things at home? [privatly to hir]
Gill — —— will vnto him
Poore: — Why, this is admirable, past my wish,
I will home instantly. nay since you will not,
Goe take hir to you, shee is your's but knowe
Your vncle and your sire shall heare of it
Gill — ——— into a di'vell
Poore — You have yor tounge at liberty, tis your owne

917 *toye*] *o* uncertain, perhaps altered; read *toyes* 920 *plight*] *l* altered, ? from *r* 921 *mee*] ^{1}*e* uncertain, perhaps altered 923 *wife*] two dots over *i* 930 *forty*] *t* lacks cross-stroke *the'are*] read *th'eare* 936 *obay*] *a* altered, ? from *e* 937 *[brin]*] heavily deleted; scribe started to write *bring* 938 *Be*] *B* preceded by a small mark like an apostrophe

B⟨u⟩t you ere long shall wish you'de tyed it vp
M^{rs} I take my leave you are revenged
The rivals doe bleed each by others sword.
Wife —— heare againe ere long
Poore I am bound to it, youngster fare you well
Keepe your word duly, or: no more but [doe] keep it.
And you my quondam betroathd, I will leave you
But knowe, the divill, will fly love as ye sea
As ships doe saile two wayes wth the same [m]winde
Soe woemen leave and take wth ye same minde

Actus 4^{ti} scæna ^{ia}

Badg: — and forsake his blewe trash — (Enter Poore
Poore: This is Quickes lodging and he [ ]hath been heere.
Badg: [———] The cheating scholler ———
Poore: — This concearnes mee much
Ime glad I heard of this, God save you S^{r}.
Badg: And you if you be worth it ——
Poore —— you have beene [wth Mr Quicke]
Wth Mr Quicke I pray yo^{u} s^{r} how fares hee
Badg: I wont tell you ——— [Fol. 36a]
Poore —— S^{r} I came from your M^{r}
Badg: My M^{r}? ——
Poore — [Yes,]Your name is Badger [is it not] e⟨n⟩t it
Bad — wth mee from my M^{r}?
Poore. S^{r} I was coming to you to this lodging
To knowe how the owner doth that if hee have

951 *B⟨u⟩t*] *u* obscured by heavy blotting that may delete another letter preceding 952 *revenged*] possibly followed by a full stop obscured by the tail of *d* in *sword* below 956 *keep*] interlined above deleted *doe* 957 *betroathd*] *a* altered 959 *winde*] *w* interlined above deleted *m* 960 *woemen*] *o* altered 961 *4ti*] altered from *3ij* 962 *(Enter*] a sweeping bracket encloses *Ent* and overlaps with *been heere* in the line below 963 *hath*] deleted ascender precedes *h* 964 initial rule inserted in error and cancelled with short vertical pen-strokes 965 dash before *This* subsequently inserted, to show that 964–5 form one line of verse 966 *you*] heavy blot over *u* 970 *I*] blot on descender extends to next line down 973 *e⟨n⟩t it*] interlined above deleted *is it not* 975 *coming*] *c* preceded by short superior vertical stroke

Required ought [b]of you from yor M^{r}, you should
Give mee the the message, you ye whilst should goe
To Medle, whom if you found dangerous
Then certifye him, Quicke is dead wherby
Hee may fly more securely
Badg: Faithfully and earnestly ?
Poore ——— As you would your selfe
Badg: ——— he doth demaund ?——— |Badg: gives him y^{e} łre
Poore: Iff I cann gett it as I hope I shall
You neede not doubt
Badg: —— then fare you well ————————
Poore ———————— Ile gull you he opens y^{e} łre.
This day is like to prove a very rare one
I never look'd for this, it came vnhoped
Fifty good pound tis well, it soundeth great
Flush in these slops; but I must not deferr.
Things falling out soe fittly I must take
All the occasions y^{t} the tymes cann make.
Exit

Actus 4^{ti} Scæna 2da

————————Sly. ——————— would it had beene a hundred. Enter
Hard ———— I am not quite cheated
Poore But you may chance to feele a new relapse
S^{r} I would speake wth you ————————
Stran⟨g⟩ ——— you may ————————
Poore ——————————————————— In privat [Fol. 36b]
Th'affaires are vrgent, Mr Quicke your freind
Commends his best love to you, wth this letter
Twill give you his full minde and his desire

977 *of*] *o* written over *b* 978 *the the*] dittography 984 *demaund*?] curved stroke of query drawn above colon *him*] *m* a minim short *łre*] cramped at edge of page; read *letter* 985 *Iff*] *ff* uncertain; could be a single letter retraced 988 *łre*] ? *r* altered 991 *Fifty*] *f* altered, ? from *s* 998 *quite*] *t* altered from *s* *cheated*] ^{1}*e* blotted 1001 *Stran⟨g⟩*] blot follows *n*; if it is top of *g*, descender has run off foot of page 1003 *affaires*] ^{1}*a* blotted; *i* written over another letter 1005 *you*] *y* preceded by curved vertical stroke

St — how fares hee S^{r}
Poore In good plight but that feare of Medle's death
Doth make him feare his life, but hee well hopes
By yor assistant love, to avoyd all
Those daungers wch as yet doe seeme to presse him
Strang ———— why came not my knave!
Poore S^{r} He intreated him to visit Medle
And learne ye hopes or feares conceivd of him.
Sly ——— fare you well good brother
Poore: Pray S^{r} commend mee to your kinsman trugull
Tell him one Change expecteth him
Sly. Is your name Change?
Poore – Yes my great man of worship
My Sly changd to a ⟨hee⟩ bosse to a swod
What, hast thou quilted thy faind gutts wth gold,
Cramb'd them wth baggs? ———
Sly ——— of my neice Gillian
Poore That was a maine one, how my Gogmagog
Sly When it is donn Ile tell you howe
Poore ———— what doubtfull?
Ney then I have out strip't thee, I did cause
Those two to fight, and for my better vengeance
Have gott this fifty pound, wch Quicke doth borrow
Of my True strange. an other hundered
[My] Gills Trugull will bring into I expect him,
And I have future hopes of ampler bootyes
W^{ch} my lawe lover, scholler hating Medle
Shall yeeld vs, I will soke him and exhaust him

1007 *that*] *a* not closed 1009 *assistant*] ^{2}*t* altered from *d* 1012 *intreated*] *ed* uncertain, perhaps *d* retraced 1017 *Change*] C written over *c* 1019 ⟨*hee*⟩] ^{2}*e* uncertain, could be *r* *swod*] ? read *sword* 1023 *maine*] the scribe did not lift the pen in moving back from *n* to dot the *i*, giving the word the appearance of *maide*, with the hook on *d* curved outwards rather than inwards 1027 *vengeance*] the word is followed by two superscript pen-marks, one resembling an apostrophe and the other a short vertical line 1030 *Gills*] interlined above deleted *My* *into*] ? read *in too* 1032 *scholler*] ? *c* written over *k*

Exantlate, pumpe out, and drawe dry his baggs [Fol. 37a]
Wee play for whole baggs wee'r no puny sharks
That venter to bee trust vp for the nipping
A bung fraught wth no more then a scotch marke
None of your Gipsyes, that prole napery
Wth shirts and smocks, no pidlers, wee doe deale
In whole sale wee, yett doe not feare a noose
A ginn to lift vs vp: lawe cann't condemne vs
To further pennance then our eares cann satisfy
Sly Tookest thou this shape?
Poore. —— to that is perfected
Revenge. but stay hee comes lett vs fall of
Stran: ———— you may tell't please it you
Poore It shall not need, S^{r} I dare trust yor word
If you'le confirme it right
Stran: Let mee inquire yor name.
Poor ———— my name is change
Sly —— as I knowe.
Poore I should have gloried to have beene admitted
Into soe grave a consanguinity
Sly And lett vs see you often ————
Poore —— I shall trouble you
Sly —— quaffe drunke wth all
Poore I take my leave
Strang To my freind
Poore ————— I shall Exit.
Actus 4^{ti} scæna 3^{ia}
Sna[]ile —— s^{r} shee is mine. Enter Poore
Med —— acquitt wth my deniall

1034 *Exantlate*] *n* written as *u*, *t*'s not crossed clearly *pumpe*] above and between ^{1}p and *u* is a vertical stroke that is possibly intended as an interlined *l* *dry*] *r* uncertain, blotted 1038 *prole*] understand *prowl* in the sense of 'pilfer' or 'filch' *napery*] *y* blotted 1039 *pidlers*] *d* altered from *l* 1040 *whole*] *l* altered from *s* 1044 the scribe began the line with a vertical descender, presumably beginning the *t* of *to*, before superimposing a long dash 1053 *a*] interlined above 1061 *Sna*[]*ile*] heavy blot or deletion follows *a* 1062 *deniall*] a long horizontal line crosses *ll*

Poore What wth a mischeif make they heere or I
This was no fitt tyme for my action
I must turne honest fate will have it soe. [Fol. 37b]
Yet Ile not loose my booty, ile attempt it
And venter gainst Ioves thunder.
Med. – may give some ayde, oh freind! ———
Poore ——— Why S^{r} your freind?
I am but will not seeme soe. your’r a villaine.
Have wrong’d a matron y^{t} deserves the stole
For hir strong chastity wth the name of bad.
Wife ——— Peace.
Poore Doe not I knowe y^{t} you did bribe ye scholler
(I have learn’t all theire trickes, & will perforce,
Despight theire pollicy turne thẽ on themselves,)
To suggest hir false to hir to credulous husband
Wth Quicke, and y^{t} [h⟨e⟩e] Quicke did outbribe him, soe
To make more easy way to worke hir false
Is not this true? deny it?
Med You dare not proove t[this].
Poore ——— oh frontlesse impudence!
What cann afford more truth to my inditement
Then his even staggering toung in his owne cause
Hee falters, faints, growes weake [ ]to excusation.
Snaile ——— receave this guilt soe pronely?
Poore Oh S^{r} sufficient reason since h’hath tried
Hir much inpregnable to all his slights
Hee would accuse hir. and no way soe strongly
As when hee would give crime vnto himselfe

1063 *heere*] ^{2}*e* altered from *a*, or vice versa 1068 *freind!*] exclamation mark has two vertical strokes converging at the bottom; resembles a *V* over a point 1070 *your’r*] ^{1}*r’* interlined above *u* and ^{2}*r* 1071 *wrong’d*] *r* uncertain, blotted, perhaps altered 1072 *bad*] sic; ? read *baud* 1075 *perforce*] *c* uncertain 1078 [*h⟨e⟩e*] *Quicke*] *Quicke* interlined above deleted *h⟨e⟩e* and *did* 1081 *t*] interlined above and between *proove* and deleted *this* 1083 *What*] entire word altered 1085 *to*] preceded by interlined and deleted superior letter or letters *excusation*] *a* obscure, perhaps altered 1088 *inpregnable*] sic 1090 *vnto*] *o* very small, blotted, obscured by descender of *y* of *way* above

Snaile ——— Then your crime was great
Poore A new vnheard of one.
Snaile. And greater love. ———
Poore It must bee soe you've wrongd them. To Med:
You must if tyme doe graunt deserve hir pardon
Med. That I may merit it
Poore No, no, you cannot
There is a death attends you will prevent it. [FOL. 38a]
Med ——— but cann't I fly it?
Poore You shall lett that suffice no signe of ioy
Snaile In that [nam]e [st⟨i⟩le] towards mee.
Poore You looke to fix'dly
Vpon this coulour, wch will dull yor sence
Of apprehension; and make mee see⟨m⟩e other
Then what I am. I yeeld I closd wth him
Why this sole end wch I did still propose
Cann give sufficient reason: my intent
Of coming hither was to free your iealousy,
To give you this chast comfort you now finde
Or elce to fix hir in perpetuall shame
Snaile I still doe thinke soe.
Poore Shall still thinke true
Whilst you continue in that fayth, inquire
Of that ill tempting scholler, if you finde him
A little differing in my maine of truth
Sepose mee from the number of your freinds
Snaile: ——— why doth hee feare death
Poore That Quicke wch caused your passion by him is not
Wife ——— much daunger may succeed
Poore: Much losse must followe I even feare to death
Med — I thanke you fare yow well

1094 *To Med:*] written within box and with rule linking it to end of next line 1101 [*nam*]*e* [*st⟨i⟩le*]] deleted *st⟨i⟩le* interlined above deleted *name* 1102 *fix'dly*] apostrophe formed continuous with tail of ascender of *d* 1104 *see⟨m⟩e*] *m* blotted 1116 *number*] *b* altered

Poore — How pretily shee doth desire his death.
But I will hope more prosperous event
Then your ill boading minde suggests to you.
For lett mee tell you, I doe knowe ye man
Cann force the rugged lawe vnbend hir browe
And fetch a smile from a more easy power;
W^{c}h shall give hir more cheerfull countenance.
Then is hir genuine, vpon faire tearmes. [Fol. 38b]
For honied speach, is an availing sacrifice;
But when a golden offring is prepar'd
You may expect not meane successe, what though
Philosophers have vrged that theire gods
Were more delighted wth ye givers minde
Then wth the glory of the haust was offred?
Yet had not men suppos'd them more accepted
They would have fitted humbler to theire altars.
Spare not a free hand & strike highest powers —
Theire sure ones y^{t} I trust to, yes soe sure
As should they wth strong hand, force man and wife
To seperation, soe to gaine a freind
A female one I meane; murder the opposers
Venter the ruine of a state, and plott
To take away competitours, they might doe it
Securely, and detected, be vnblam'd
Att least vnpunished

Med — much easily obtained ————

Poore — Wth out much difficulty
But you must thinke y^{t} in externe affaires
Theile not soe strongly labour wthout hope
Of future benefitt.

Med — —— blood and spirit away?

1129 *vpon*] *v* altered, ? from *f*, anticipating *faire* 1130 *speach*] *c* altered *sacrifice*] *e* blotted 1135 *haust*] an unrecorded spelling of *host*, in the sense of 'sacrificial offering' (Latin *hostia*) 1142 *I*] altered

Poore Your life I will secure mee on myne owne
If wee conclude agreement for what summ
Med Being your creature
Poore S^{r} prepare the summ
Against I bring you life ——
Med —— I shall, what is it?
Poore An easy one I dare venter it for 200
Med ——— vnlesse my tombe [FOL. 39a]
Poore: These sacred meditations strongly fitt
Men given to observance of true virtue;
But thinke not only, of your last good S^{r}.
For there are many mediates wch require
Some like respect wth that.
Med Who have long hope to escape that. ——
Poore [] ————— then wth you
For heere is that will give you lives assurance
For this crime
Med Have you a pardon S^{r}?
Poore ——— probatum est
And S^{r} wthout compelling articles
Your will is theire desire, what you shall please
Wilbe sufficient vnto thẽ acceptedly.
Med ——— a deniere from it
Poore Your hand wilbe to liberall, they procurd it
Wth a small easy breath.
Med And then at last hardly obtainde. ——
Poore ————— tis true
I will accept for them, what you shall please
Med ——— and Ile deliver it
Poore I will, the waight of my deserts, how strong
It is how forcible this benefitt?

1161 *meditations*] interlined above caret; ^{2}*i* blotted 1167 single vertical stroke, probably descender of *T*, deleted at beginning of line 1173 *will*] *ll* uncertain, perhaps altered from *tt* 1174 *Wilbe*] vertical stroke precedes and runs into *W*; *lb* joined by ligature

When should his coyne bee wth my pardon layed
In a true ballance myne would bee outwaigh'd,
Tost in to aire; What I receave I gett
Giving him for his sterling counterfett
Wth w^{ch} [hee]sIme well appayde, hee is well pleasd
Hee that hath to much may of some be eas⟨ed⟩
Exit

Actus 5^{ti} scæna i^{a} [FOL. 39b]

Badg: — Slid heare comes somebody Ente Poore.
Badg. you shal bee mett wth S^{r}
Poore I must now doffe this covert of my villainye
Quicke I must thanke thee for thy words have been
An ample gaine to mee, and Badger to
Badg: A sees mee not trowe, doth hee?
Poore ——— thou hast binn
A great ayde to mee, I must give thee thanks.
Badg: ——— when you knowe all.
Poore How evesdropt
Badg — Hee [] hath not the same beard.
Poore Ile wash and shave you, and yor greasy blewcoat
My serving ⟨d⟩onn I will; but I must forgoe
This fifty pound now I am caught wth it.
Twill make a deepe hole in my summs, a la⟨n⟩ke
W^{ch} all my letting out cann nere make full.
I would some taylour would instruct me fairly
To patch vp this misshapen sute againe
And give it wthout bracke. Well I [will keep] le not loose it.
But yett to loose my vncle were worse ill

1185 *outwaigh'd*] *d* perhaps altered from *t* 1188 [*hee*]*sIme*] *s* should also have been deleted; an apparent ligature between *s* and *I* is the loop beginning *I* 1189 *eas⟨ed⟩*] the ending of the word is difficult to decipher, but seems to be *ed*, despite the rhyme with *pleasd* in 1188; a descender is visible between the two letters 1192 *Ente*] sic 1194 *villainye*] *ai* uncertain, blotted, perhaps altered 1196 *Badger*] *B* altered from minuscule 1202 *hath*] preceded by single deleted letter, possibly *t* 1204 *⟨d⟩onn*] *⟨d⟩* uncertain, possibly badly made *D* 1206 *la⟨n⟩ke*] *⟨n⟩* perhaps altered, ? from *u* 1207 *make*] *k* uncertain, perhaps altered 1210 *le not loose*] interlined above deleted *will keep*; thus *Ile* coheres 1211 *But*] preceded by a short dash in margin

Let it prove how it will Ile venter it
Abide the hazard of it, Ile tosse fairly
To scape, fortune must be my opposite
If I doe loose it.
Badg: A mischeife on your muttering chops
Poore ———— Have at you
I left it heare, and I must search it out.
Badg ——— but not soe well
Poore True for the savour's worse.
Badg As thinn a roome as may bee
Poore: –I remember
Twas on this side I layd it; what have I heare
What is it turn'd into a baskett hilt
And threadbare blewe coate, twas a good exchange [FOL. 40a]
For him that made it; vm, may not the snake
That cast the skinn be found heare, nor ought elce?
Nay Ile search furder; oh you miching raskall
What have I found you? You shall pay for it.
The raskall was crept vp into a mouse hole
And lay as close as a hedge hogg: what freind Badger?
Badg: I even the same S[r] ————
Poore What makst thou heare now?
Badg –And. —— and ——.
Poore What, what then?
Badg —— [you doe knowe his humour] And I dare not venter.
Poore What.
Badg till his anger's past.
Poore Tis well, were not thy parents puritanes?
Badg: [W] —— why doe you aske
Poore Did they not teach thee for to pray extempore
Badg But when they went to them
Poore ———— did they not hum and ha
When they were gravelld

1228 *miching*] ? *h* altered 1236 *And . . . venter.*] interlined above deleted *you . . . humour*

Badg —— yes perchance they did
Poore And when thou wert gott ———
Badg: I don't remember that
Poore Mee thinks they should it seem's innate to thee.
But thou'st reduced it better to thy art
Of lying; I doe knowe your busines mungrill
Your sett to spy my noble trencher man
You've waited all this while but for small cheare
An howers attendance had beene better giv'n
For but a head of garlicke, see you this steele?
Ile make you munch a peice of't if y^{u} swear not [FOL. 40b]
As I shall vrge, but if you sweare looke heare
Crounes you mad raskall.
Badg Then I will sweare
Poore —— tis well sayed but this place
Is no fitt one for quarrels, will you sweare?
Bad: Since I am forced I will
Poore ———— thou shalt no furder
Then I allready have: you shall conceale mee.
Not give him notice, that I was ye factour
Who tooke vp fifty pound on Quicks behalfe
Badg: Why by this hand I wont.
Poore What doe you equivocate
And sweare by your leffe hand whẽ you mean to write it?
Sweare you by both your hands
Badg: —— by my both hands
Poore — nor either of them.
Badg [ne] —— neither
Poore Nor your tounge
In word or signe you shall make any way:

1253 *giv'n*] *v'* uncertain; *v* closed and slight blot runs into possible apostrophe 1254 *steele*] ^{2}*e* blotted 1255 *not*] possibly written in pencil 1257 *raskall*] ^{2}*a* not closed, could possibly be *e* 1263 *allready*] vertical stroke runs into ^{1}*a* 1265 *behalfe*] *l* altered from *f* 1267 *doe*] *d* blotted 1268 *leffe*] an unrecorded spelling of *left* *write it?*] interlined within bracket above *mean to* 1273 *tounge*] short horizontal line under *u* 1274 *way:*] faint upward line above colon

Badger No way by signes or tokens
Poore —— this thou swearst
By thy sword hilts, for thats the hardest oath
I cann now force thee to.
Badg ——— I doe sweare this.
Poore Wthout reservances
Badg I from my heart
Poore Then heare my noble skincker heare is gold
Twill give thee freise in stead of thy blew coate.
Twill give thee gaudyes, thou mayst cram̃ thyselfe
Wth kicksh⟨a⟩wes now, as long as this shall last
Whilest this resplendant substance shall remaine
Wthin ye repleat body of thy purse.
This hath sufficient spirit, centinell. [FOL. 41a]
Twill give thee douszens, more then perfect summs
They shall exceed the prædicament's best number
And the 3 principals: three shall not bee all.
Twill make thee looke, like a Claridiano
Till it hath made thee a hebitated Zoophyton.
Badg ————— wth your coniuring tearmes
Poore Fare well good badger, I have other busines
I should bee more intent to.
Exit.

Actus 5^{ti} scæ 2da (Poore sitts at his (study

Sly —— and ready in that art, I would faine h⟨ear⟩e him ——
Stran Heare a lector from you
Poore Most willingly though Ime not ⟨we⟩ll provided.
Sly. Wee will expect the lesse
Poore ———— Ile give you breifly
The texture of a speechfull composition.

1282 *skincker*] [1]*k* blotted, could be *ck* 1283 *stead*] *a* uncertain, blotted 1293 *hath*] *t* blotted 1298–9 s.d. in two rounded brackets 1300 *art,*] the comma is dropped below the line, and meets the *y* of *you* in 1301 *him*] *m* a minim short 1302 ⟨*we*⟩*ll*] dot over [1]*l*

When the infernall h⟨e⟩lbread shades of night
The hate of Phœbus, and the scorne of light
A're forc'd to theire darke cells, choyce spirits arise
From theire dull easyes frightlesse lethargyes.
My spirits are not fresh, the subiect's mourning
Aurora wane, first the etymology
The golden hower, when Phœbus first displayes
Vnto the ioyed world his more ioyfull rayes
Now amplyfy it frõ the propertyes
Extract's the vapours, from the thickned aire
Expels' the sadnes, gives it subtile, rare.
The effects doe followe wch our bodyes have
And wch our minds, externe and interne these [Fol. 41b]
Our blood our nerves receive like purity
That from the aire, wee from the purged sky
Should we dampd [aires] nights polluted aire still breath
As wee receivd life wee should drawe in death.
But being cleansed by that sacred fyre
That aire feeds life blest life, our best desire
Now for the operation in our minds.
What ofspring of high witt, birth of rare art
W^{ch} from this tyme doth not acquire cheife part.
I should proceede to prove this by connexion
The mourning salutations were calld holy
Amongst the Romans. then wee may surmise
Those studyes holy that wth Sol doe rise.
For then there is a greater sympathy
Betwixt the stars and vs, they stand more nye
To eloquence, and helpe more o^{r} theorie.

1306 *h⟨e⟩lbread*] ^{1}*e* blotted 1308 *A're*] *re* uncertain, faint 1309 *lethargyes*] ^{1}*e* uncertain, blotted, obscured by descender of *s* of *spirits* above 1311 *wane*] *w* uncertain, could be *m* 1313 *Vnto*] *V* altered 1317 *followe*] *ow* blotted 1319 ^{2}*our*] *o* altered, ? from *b* *like*] ? *k* written over *f* 1321 *nights*] interlined above deleted *aires* 1326 *witt*] *w* altered, ? from *b* anticipating *b* of *birth* 1330 *Romans*] *o* blotted 1334 *o^{r}*] *o* uncertain, resembles closed *v*

Now should be some proportioned inductions
To prove that tyme most apt to meditation.
Then follow individuall examples
Of such as have vsed it these must be sett downe
In grave words, full and sounding; well connected
Agreeing in theire sence, and these not vulgar.
Hyperb⟨o⟩lyes sometymes, then Metaphors
These now wthout coniunction, though not often,
Yet bearing still relation on, to other.
Now vse an iteration, speake w⟨o⟩rds twice,
But lett them still bee increasing, and ascend
Not falle to flatly. soe heare are instructions
Such as the tyme, and my weake braine cann give [Fol. 42a]
Quicke – how to compose a speech
Poore Not any⟨-⟩one
As I remember doth sett these downe fully.
Some heare some theare, I have collected, not
sucking my hony from one only flower,
But From [the] best fountaines Aristotles rheth'ricke
Tully in 'his oratory, from Quintilian.
Badg ——————— doe you meane
Poore [E] ———— No badger no .
Ex M Fabij Quintiliani institutionibus.
Badg ——————— by fifty pound.
Poore: Yes S^{r} some fewe affaires calld mee abroad
And force'd mee bee lesse diligent, then I would
But now theire ended, I shall give attendance

1335 *inductions*] ^{2}*i* has extra loop following it, probably the result of failure to lift the pen before making the dot. The same feature occurs in ^{2}*i* of *meditation* at 1336 and ^{2}*i* of *iteration* at 1344 1340 *vulgar*] *r* altered, perhaps from *t* 1343 *on, to*] written as one word with the comma presumably placed to indicate separation *other*] *h* uncertain, retraced 1344 *words*] *o* blotted 1347 *give*] *i* has double dot; smudged blot partly obscures *e* 1348 *speech*] ^{2}*e* blotted, perhaps altered from *a* 1349 *any-one*] erroneous descender precedes *y*; hyphen doubtful 1351 *collected*] ^{2}*c* altered from *d* 1353 *But*] interlined above *From* *rheth'ricke*] ^{2}*h* may be altered, ? from *o*; apostrophe uncertain, formed as superior elongated loop 1356 [*E*]] letter probably written as result of eye-skip to *Ex* on 1357 1358 *pound*] *d* written over *t* 1360 *force'd*] the apostrophe may just be the detached ending to the loop on *d* *diligent*] a line rises from *n*

More amply to you.
Quicke ——————— tis best.
Poore that's the best way to thrift [indeed] where is your neice
Sly ——— a dodkinn wth my will.
Poore You were to much obdurat then, to hard
You may spoile all hir possibilityes
Such great extreames force naught but desperatiõ
Quicke ————— for your great labours
Poore S^{r} if my best indeavours could deserve them
I should account them, very strong reward.
S^{r} my desire of gaine is not soe stupid
As is your common pedants, yet no ambition
Hath grow'n soe much [vp] on mee as I should covett
A meerely nominall opinion
Oh affectation is a cloudy vayle
W^{ch} hidst the solidst, of our soules perfections. [Fol. 42b]
Or at the least doth hinder hir free workings
Quic: [—————] of your free soule ———
Poore S^{r} I proffesse, an essence
W^{ch} should as perfectly bee knowne as bee.
But since the wretched, vild esteeme of men.
Doth give the best of men but meere selfe lovers
If they esteeme them selves, I gratulate
Your good coniecture, that you thinke mee free
Whilst I doe knowe my selfe soe, fare you well S^{r}.
Stran ——— anon Ile make a second visitation
Poore You may expect mee ready to yor vowes
Badg: ———— since to day.

1362 *you*] *ou* uncertain, blotted, with what appears an apostrophe over *o* 1364 *where*] interlined above deleted *indeed* 1366 *obdurat*] *ob* blotted, perhaps altered; a dot is visible above *o* 1368 *desperatiõ*] tittle uncertain, formed continuous with *o* at very edge of page 1370 *deserve*] *s* altered from *d* 1373 *ambition*] *b* strangely formed, bottom loop resembling top loop of *p* 1379 initial rule wrongly inserted, then deleted with wavy line (1379–80 form one line of verse) *soule*] *e* altered from *d* 1382 *esteeme*] additional descender follows *t*, sloping so as to join *t* at the bottom 1383 *best*] *s* uncertain, blotted, perhaps altered 1384 *selves*] *v* altered from *f* (cf. 1433) 1387 *second*] blot before *d*

Poore — Yes Badger if thoult give mee ample thanks
That I've remembred thee soe well.
Badg — ———— oh Lord S^{r}!
Poore — Soe now they'are gonn what wouldst thou my brave pufpast.
What wouldst thou wullsacke, whose inside is no better,
Then 'a sheeps coate, ift bee of equall goodnesse
Sly — my wandring prince of troy. ————
Poore: — ———— why thou shallt knowe
I will rehearse my ephemerydes
Myy dayly slights, since moondayes last meridies
But thou must bee my subiect and my scænicke
To act my gulls in glorious wise.
Sly — ———— content
Poore — Weele first beginn wth strange
Sly — ———— Heare I come
Poore — sound tr⟨u⟩mpetts heere our play begi⟨nne⟩s
Sly — and vitiated your muse.
Poore — ———— fy thou art out
I am his true begott, legitimate.
Sly — —— ⟨b⟩y making pallinodes [FOL. 43a]
Poore — And thou wouldst live soe to, well Ile instruct thee.
Sly — I would. ————
Poore: — ———— but first you must putt of your fatnesse.
Pooets are leane and marc⟨e⟩lent
Sly — ———— hir burden dead
Poore. — Well thought of, oh I have the finest lasse
Have made the bravest conquest, purchase of hir.
I hope none heare⟨s⟩ Ile tell thee shee excels

1393 *thou*] *u* uncertain, perhaps altered *pufpast.*] interlined within bracket above *my brave* 1394 *wouldst*] *ul* uncertain, perhaps altered *wullsacke*] *u* uncertain, perhaps altered *better,*] ? a very short dash after the comma 1395 *Then*] *n* appears to be written over a second *e* *equall goodnesse*] written as one word *goodnesse*] ^{2}*e* altered or retraced 1397 *shallt*] long and short *s* superimposed *knowe*] heavy blot under *w* 1399 *Myy*] sic *meridies*] ^{2}*e* altered or retraced 1401 *glorious*] *s* uncertain 1405 *tr⟨u⟩mpetts*] ⟨*u*⟩ very small, blotted, could be *o* *heere*] ^{2}*e* uncertain, could be *a* or altered from or to *a* *begi⟨nne⟩s*] *g* and *nne* blotted, perhaps altered 1413 *Pooets*] *oo* uncertain; ^{1}*o* not closed, and runs directly into ^{2}*o* *marc⟨e⟩lent*] ^{1}*e* uncertain, possibly undotted *i*; read *macilent*

Man in's best property of looking vpwards
Hir falling eyes give heaven full viewe.
Sly — no more deserving qualityes
Poore: Such as your common women have shee's coy
Yet wanton, shee cann laugh, and weepe, and laugh,
And hould againe. shee hath an exquisite face
And yet not painted wch is very rare.
Sly —— transccends shee Gill?
Poore —————— yes fayth in feature
But Gill hir more in witt and haviour.
And heere shee comes; what may ⟨wee⟩ wish y^{u} ioy
Of your good match?
Gill That did indeavour cousenage
Poore ——— How, certainly?
Sly — a rocke quite shipwrackt
Poor It cannot bee, none knowes vs but our selves
And wee o^{r} selves soe finally, as no humour
Could give mee knowne vnlesse yor womans tounge.
Yet Ive one refuge and, it is my last
The very sanctuary of our safety
As I supposse it yet, but prove that wanting [FOL. 43b]
I cannot guesse the consequent save ill
Sly Lets know't
Poore [] —— ney much of ill must force y^{t} from mee.
Gill And suffer a small bafling
Poore ——— oh I cannot
But why what proiect, what event will followe?
Gill I have revealed your disguise.
Poore ——————————— how, how?

1419 *Hir*] *r* blotted 1421 *coy*] *o* blotted 1423 *hould*] *u* interlined above *o*; *d* altered from *e* (i.e. *hole* altered to *hould*) 1428 ⟨*wee*⟩] letters very heavily blotted 1429 *match?*] extra vertical line following question mark 1432 *shipwrackt*] *t* altered from *d* or vice versa 1433 *selves*] *v* altered from *f* (cf. 1384) 1434 *finally*] *y* has two descenders 1438 *supposse*] *se* blotted 1441 single vertical stroke deleted at beginning of line 1446 ¹*how,*] short, thick line above comma may be intended as exclamation mark

Ime tangled in a cobweb that have scapd
Snaires and strong engines able to prevaile
Against a lion, if the fox were absent
But now the ridle is confirm'd, a secret
Is much to little for one only man.
For two sufficient, but for three to much.
Well goe thy wayes, old Gill, Ive knowne thy equals
But bedlam kept them for they could not themselvs
Wthin due compasse is your Trugull heare

Sly [—] what new shape may I take ————
Poore Why turne a horse leech.
Thou mayst sucke blood securely in y^{t} habit
Somewhat Ile doe and labour for event
W^{ch} shall alone give knowledge what I meant.

Exit Actus 5^{ti} scæna 5ta

Sly ———— houle like sterved currs.
Poore For mee I am the obiect, may they burst.
Conceale mee lett mee not bee knowne.
Sly ———— knowe you not Quick's death?
Poore Oh y^{t} nothing moves mee, I divulg'd him dead
For my owne private ends
Sly And h⟨e⟩e is dead
Poore. Poets are prophets then I see! how dead?
Amasement ceaseth mee, dead? it cannot bee.
Why then a necke verse followes, oh my fate [FOL. 44a]
Woemenes best witt I see is extreame folly
How free[] had I beene from this certaine ruine
How practis'd in more ills, had not this hapned
And flourishing in them? ist not possible

1455 *compasse*] *a* uncertain, could be *e* 1456 initial rule wrongly inserted, then deleted with three vertical pen-strokes 1459 *event*] second upstroke of *n* written over longer upstroke, presumably premature beginning of *t* 1466 *dead*] *e* blotted, *a* interlined above and between *e* and *d* 1469 *prophets*] ^{2}*p* preceded by (perhaps altered from) single vertical stroke *see!*] punctuation uncertain; either exclamation mark or query appears to have been added to comma 1470 *Amasement*] *A* altered, ? from *C* 1471 *oh*] the ascender of *h* is very faint and short 1472 *Woemenes*] illegible stroke follows *n* 1473 *free*] single letter deleted following ^{2}*e* 1474 *had*] ? *h* written over *a* 1475 *possible*] ^{2}*s* retraced

That I may live vnknowne to Medle? tis
And I will venter it, shake of these burrs
Wth easy recompence of a little nap.
You shalbee a phisition, I am sicke
You make me daungerously sicke, but heare you
Ile not bee purged, you shall give me out [p⟨ur⟩g'd] sicke
But not give inward sicknes, Ile no figgs.

Sly. As you shall please.

Poore ——— I doe not like[, I doe no] the humour
Of your great guilty person[ages]s, who to scape
A lawfull death; that is death giv'n by'th lawe
Will rather choose to dy, vnnaturally
By theire owne guilty hands.

Sly ——— wth mature iudgement.

Poore Intreat Strange hither

Sly ——— to prevent my labour?

Poore ——————— fittly.

Strange ——— termes wth one consent

Poore S^{r} lett mee crave your pardon, I esteeme you
A second parent to mee, removed by nature
But one degree from it, you are my vncle
I therfore will lay ope my worst acts to you,
That you may veiwe them fully, as they are
In theire owne essence: I have wrongd them all
And giv'ne iust cause for this complaint, nay more
W^{ch} most afflicteth mee, I have wrongd you

[BLANK] [FOL. 44b]

[BLANK] [FOL. 45a]

Strange by wch you did conceive mee soe. [FOL. 45b]

1477 *venter*] ? *n* written over *r*; mark like apostrophe to right of *n* 1481 *sicke*] interlined above deleted *p⟨ur⟩g'd* 1485 *person[ages]s*] ^{2}*s* interlined above deleted *a* 1490 *Strange*] *t* altered from superior *r* 1501 *you*] blot follows, conceivably deletion (? of superior *r*)

Poore — I shalbee knowne
sufficiently heare after.
Sly And putt it in to practise: —
Poore ——— I doe promise
A like restraint from the vnciv⟨i⟩ll liberty
Tyme and our ryoutous age doth prompt vs to
Str: ——— choakd wth recompence
Poore Wee are deficient in ability.
Sly ——— stop'd till cramm'd
Poore Since the whole summ of my continued actions
Have been me⟨'⟩re tricks, Ile end them wth a tricke
Ime sicke to death.
Strang [———] the reast Ile vndertake ———
Poore ——— Let them fly in
Give mee a gowne and nightcapp
Sly ——— heare they are.
Poore Wheres your phisitions habits, have y^{u} termes.
Fustian will serve sufficiently curiosity
Will stand you in no steed, heere are no Criticks
Stran: ——— Ile admitt them
Poore S^{r} I am ready for them, for some meale now
To make a wh[ighte]ite man of mee & a sickly.
Oh, oh, oh.
Sly ——— whats the disease?
Poore: ——— The epylepsye
Sly The falling sicknes?
Poore ——— I ———
Sly And much good doe it you.

1507 *vnciv⟨i⟩ll*] ^{1}*v* and ⟨*i*⟩ blotted; lacking a dot, ^{2}*i* might have been intended as *e* 1512 *whole*] *l* altered from *s* 1513 *me⟨'⟩re*] apostrophe obscured by descender of altered *s* in *whole* at 1512; possibly a second *e* is intended 1515 initial rule wrongly inserted, then deleted with two vertical pen-strokes 1519 *have*] *a* blotted 1524 *make*] *k* written over *s* *wh[ighte]ite*] *ite* interlined above deleted *ighte* 1525 ^{2}*oh*, ^{3}*oh*] both *h*'s uncertain, perhaps altered 1527 *epylepsye*] ^{1}*y* altered from *i* 1528 *sicknes*] *k* altered, descender (? of *y*) visible; *n* faint, *es* retraced 1530 *much*] *c* written over a letter with an ascender

Poore — I hope it will
Sly How didst thou knowe him for thy vncle?
Poore ———— Strangly
Some other tyme ile tell you; they are entring.
Tru: Made mee a gull. [Fol. 46a]
Poore Oh, oh, oh, I confesse
That, [yo]u I have beene the cause, youve suffred wrong
Dry —— agree to it. (— shee gives him gold.
Poore: Ime heartily sory for it, I thanke my god.
He []hatth brought you hither, that I may crave | hee
(falls downe
Your pardons, I would my estate were able (in his fitt.
Sly — present at, how cheare you?
Poore Why well I thancke my maker, fitt for heaven
If these could be intreated to forgivenes.
The remnants of what I have gott from you
I will restore wth thanks to satisfy you
Stran ——— that Ile not vndertake
Poore I thank[] you your carefull in my behalfe
Stran In presence of these gentlemen.
Poore ——— there is one absent
One Mr Medle, him I would faine speake wth
Str⟨an⟩g — whom you desirde to speake wth
Poore —— I must intreat
Your pardon for Ive wrongd you.
Med Hard Tru: —— wee doe to wch beare witnesse
Poore Then thus I shake my sickenes of
[Trugull I for my loving spouse].
[Poore happily may you live.]
Med: why did you crave my pardon?

1534 *they*] *e* altered 1537 *I*] interlined above [*yo*]*u* 1540 *hatth*] ? adjustment of *that*; ${}^{1}h$ preceded by deletion, ? of *t* 1542 *fitt*] *tt* heavily blotted 1545 *forgivenes*] *s* uncertain 1549 *thank*] deleted or heavily blotted single letter, possibly *e*, following *k* 1553 *Str⟨an⟩g*] ⟨*an*⟩ heavily blotted 1558 [*Trugull*]] *Tru* altered from *Med* 1559 if line was correctly allocated to Poore, deletion may be a theatrical cut

Poore — [But ⟨wha⟩] — 'S^{r} I craved
But what I gave you, doe you knowe mee now?
I am to all of you what you will but good.
Med — Is then my pardon counterfett?
Poore — ——— twas the best
That I could give you; Ive no more from you
Only the difference is I payd not for it [Fol. 46b]
An equall price.
Med. — weele both have equall parts; ———
Poore — ——— tis fairely offred
Sly — All thrive but my selfe.
Poore — My gaine is thine; for what remaines in bank
Of our last getting shall restore thy state.
And give thee means of trading, one ill fate
Wee equally indured, fortunes sad frowne
Wee shared betwixt vs, but it is my croune
That as in worst of ill thou hadst a p̱t
Soe of our [better]best state thou a sharer art
This is the maine true freindship cann com[m⟨aun⟩]maund
Y^{t} hopes and fears of freinds goe hand in hand

1561 *Poore*] *e* blotted by deletion following [*But* ⟨*wha*⟩]] cf. 1562 1569 *Med.*] *M* altered 1570 *Poore*] set to the right because of an indecipherable mark in the margin 1574 change to coarser pen, letters tend to blot 1576 *shared*] *re* uncertain, run together *betwixt*] *w* a minim short *croune*] *e* uncertain 1577 *p̱t*] scribe running out of space; read *part* 1578 *art*] interlined above *sharer* at edge of page 1579 *com*[*m*⟨*aun*⟩]*maund*] *com* runs to binding, with deleted *m*⟨*aun*⟩ interlined above *com*; *maund*, presumably a second attempt, is interlined above *freindship cann*

MUSICAL AND DRAMATIC DOCUMENTS FROM THE MIDDLE TEMPLE

These records of musical and dramatic entertainments were prepared from documents at the Middle Temple by John R. Elliott, Jr. and checked by H. R. Woudhuysen.

November 1992 N. W. Bawcutt

In Malone Society *Collections XII* Tucker Orbison edited a number of previously unpublished documents concerning the productions at court of George Chapman's *The Memorable Masque . . . of the Middle Temple and Lincoln's Inn* in 1613 and of James Shirley's *The Triumph of Peace* in 1634.[1] These documents have survived in the archives of the Middle Temple because members of that society acted in each case as the general treasurers for the productions, which were sponsored by all four of the Inns of Court. The documents printed by Orbison had been gathered together by an archivist in the early 1970s and placed in a cardboard carton labelled 'Masques and Entertainments'.[2] At the same time, however, various other relevant documents were sorted and placed in six further cartons, arranged chronologically and, within each carton, by type of document. Some of these supplement the records of the 1613 and 1634 masques, while others refer to further Middle Temple entertainments during the first half of the seventeenth century. By kind permission of the Masters of the Bench of the Honourable Society of the Middle Temple, transcriptions are given here of the entries that may be of interest to musical and theatre historians, together with brief descriptions of their context.

1. Documents Relating to Chapman's *The Memorable Masque*

1. A. Loan receipt dated 25 January 1612/13, recording a loan of £50 from Francis Moore towards the expenses of the masque.

25° Die Ianuarij 1612
Receaved of ffrauncis more Esquier on of the m^{rs} of the benche of the middell Temple to the vse of the house the som of fifty pound*es*. I say Receaved
R Daston

This was the first of several loans made to the society by senior members before a taxation could be levied on the rest of its members. Moore had recently been elected a Master of the Bench in 1612, and became Serjeant-at-Law in 1614. His loan is also recorded in Orbison, document C. 11. Richard Daston, also

[1] 'The Middle Temple Documents Relating to George Chapman's *The Memorable Masque*' and 'The Middle Temple Documents Relating to James Shirley's *The Triumph of Peace*', ed. Tucker Orbison, Malone Society *Collections XII* (1983), 1–84.

[2] Orbison, 'Research Opportunities at the Inns of Court', *Research Opportunities in Renaissance Drama*, xx (1977), 27–33.

a Middle Temple Bencher, served as the general treasurer for Chapman's masque.

1. B. Loan receipt dated 24 February 1612/13, recording a loan of 100 marks, or £66 13*s*. 4*d*., from Sir Henry Mountagu towards the expenses of the masque. Compare Orbison C. 3, a memorandum written sometime later, which dates the transaction 'about ye xvjth' of February.

xxiiij^{to} die ffebruarij Anno 1612
Receaved the daye and yeare above written of the right wor^{ll} S^{r} henry Mountague Knight Recorder of the Cittye of London to the vse of the ffellowshipp of the Middle Temple the some of one C mark*es* Curraunt money And, y^{e} sayd some \of/ one hundred mark*es* ys to be repayed to the sayd S^{r} henry Mountague his executors or Assignes so soone as the sayd some maye be convenyently levyed of the ffellowshipp of the Middle Temple

Receaved by me Richard
Baldwin vnder Trer of the
Middle Temple.

Sir Henry Mountagu of Boughton, later the first Earl of Manchester and Lord Chief Justice of the King's Bench, was the Recorder of London in 1613. His two elder brothers, Sir Edward and Sir Sydney, were also members of the Middle Temple and took an active interest in the masque (see Orbison, pp. 6, 28–9). Richard Baldwin was Under-Treasurer of the Middle Temple from 1591 to 1619.

1. C. Certificate dated 21 February 1612/13, admitting Inigo Jones, who designed the scenery for the masque, as an honorary member of the Middle Temple. Endorsed: 'Iones Inigo ar' et Trowbridg g: ar' ad̄ p*er* lectorem gratis 21 die febr' 1612 4 et 5'.

xxj^{mo} die ffebruarij 1612
Inigo Iones ar' admissus est in Societatem medij Templi Specialiter ex assen*su* Rob*er*ti Davies ar' mod lector' et alior*um* M$\overline{ro}$*rum* de Banco

Jones's admission was also entered in the Middle Temple Parliament Book on the same day.[3] Jones's sponsor, Robert Davies, had been a member of the bar since 1594, and was involved in auditing the accounts for the masque.

1. D. Orders of the Middle Temple Parliament, dated 11, 16, and 18 June 1613, concerning the financing of Chapman's masque. Complete transcriptions are given here since the abridgements printed by Charles Trice Martin in his edition of *The Middle Temple Records* are sometimes inaccurate and misleading

[3] *Middle Temple Records: Minutes of Parliament of the Middle Temple*, ed. Charles Trice Martin, 4 vols. (London, 1904–5), ii. 561.

(see, for example, 1. D. 4. and commentary, below). The texts are taken from the 'Parliament Book 1610–26' in the Middle Temple archives (no shelf-mark).

1. D. 1. 11 June 1613. (Fol. 54a)

fforasmvch as at this tyme the howse standeth indebted in dyvers somes of money amounting in the whole to the some of six hundred pound*es* and moore, wheareof a greate parte of the said some of six hundred pound*es* ys owing vnto dyvers artificers in this cittye, such as are not able to forbeare their money but to their greate losse or vndoing, ffor avoyding of all such inconveniences wch maye heereafter grow by reason of the said debt, The ma^{rs} of the Bench have at this present p*ar*liam^{t} decreed and so ordered that there shalbe forth wth anoth^{r} taxacion presently levyed thorough owt this howse, vidlt everye one of the ma^{rs} of the Bench to paye iij^{li}, and everye one of the ma^{rs} of the vtterbarr to paye xls, and everye othr gent' of this ffellowshipp to paye xxs Provided that all such gent' of this howse that have bynn admytted since the first daye of Easter Tearme last, and all others to be admitted heereafter shalbe free from this tax./

1. D. 2. 16 June 1613. (Fol. 54b)

Whereas at the p*ar*liam^{t} holden the xjth daye of this instant moneth of Iune, yt was then ordered by all the ma^{rs} of the Bench, that there showld be an othr taxacion presently levyed thorowghowt this howse vidlt every one of the ma^{rs} of the Bench to paye iij^{li}, and every one of the ma^{rs} of the vtterbarr to paye xls, and every other gent' of this ffellowshipp to paye xxs, Synce which tyme the ma^{rs} of the Bench vnderstanding the vnwillingnes of the gent' to be taxed at such greate somes, and the saide ma^{rs} of the Bench willing to ease them as mvche as in them lyeth, have ordered that whereas the ma^{rs} of the Bench weare taxed at iij^{li} a peece, to be now but ls apeece which ys forthwith to be payde, and everye vtterbarrister which was to paye xls ys now to paye but xxxs the one half thereof to be payde before sundaye next, and the other half to be payde before the ffeastdaye of all Saynt*es* next commyng, And every other gent' of this ffellowshipp which was to paye xxs a peece ys nowe to paye but xvs the one half thereof to be payde before Sundaye next and the other half to be payde before the ffeast daye of all Saynt*es* next commyng as aforesayde, and all gent' that are Attorneys or practisers of this howse to paye xxs a peece, the one half thereof to be payde before Sundaye next and the other half thereof to be payde before the ffeast of all Saynt*es* next commyng as aforesayde, And further whereas dyvers gent' of this ffellowshipp have vsually come into Commons in the vacacion tymes, and not in the Tearme tymes onelye to avoyde the payment of duetyes, yt ys therefore further ordered that all such gent' that now are in Commons, or that shall come into Commons at anye tyme heereafter, shall not goe owt of Commons before they have payde all such some or somes by them due as aforesayde./

It would seem that no attempt to collect the higher rate was ever made, for a document of 12 June 1613 (Orbison, C. 9) already shows the lesser rate in force. A still smaller rate had been applied in the first taxation of 11 January 1612/13 (Orbison, C. 1), which was not recorded in the Minutes of Parliament.

1. D. 3. 18 June 1613. (Fol. 55a)

Whereas Edward Powell gent' one of the ffellowes of this howse hathe of his voluntary mynde lent vnto the ffellowshipp of this howse toward*es* the defraying of the chardg of the late maske the some of fiftye pound*es* currant money, ffor repayment whereof the ma^{rs} of the Benche have at this present p*ar*liam^{t} with one consent decreed and so ordered that the saide some of fyftye pound*es* shalbe repayde vnto the saide Edward Powell his executors or assignes at the ende of six monethes next after the date of this present p*ar*liament./

The purpose of this order was to extend the period of the loan, the repayment having originally been stipulated for six months from the date of the loan, which was 18 February 1612/13 (Orbison, C. 4). In fact, however, Powell received his money on 28 June 1613 (Orbison, C. 28). Edward Powell had recently been admitted to the bar at the Middle Temple in 1611.

1. D. 4. 18 June 1613. (Fol. 55b)

And further wheareas Richard Baldwin, ffrauncis Barker, & Arthur Blunt, officers of this howse, at ye instance of the right wor^{ll}: the ma^{rs} of the Bench of this howse, doe stand ioyntly and seu'ally bound in one obligacion bearing date the xvth daye of ffebruary last past vnto Sr Thomas Temple of Stauntonbury in the Countye of Buck knight in the some of fower hundred pound*es* of lawfull money of England, with condicion for the paym^{t} of two hundred and fifteene pound*es* one the xviiijth daye of November next ensuing, as by the same obligacion and condicion more at large yt doth and maye appeare, which money was imployed toward*es* the defraying of the charge of the late maske, Now the ma^{rs} of the Bench have at this present p*ar*liam^{t} with one consent and agreem^{t}, decreed & soe ordered that the saide som*me* of CCxv^{li} shalbe repayde vnto the sayde Sr Thomas Temple his executors or assignes one the saide xviijth daye of November and according to the tenor and effect of the sayde obligacion & condicion,

Martin's transcript of this entry omits the figure of £400 as the amount of the loan (*Minutes of Parliament*, ii. 566–7), thus erroneously giving the impression that the repayment was for the total amount rather than an instalment of it. Orbison accordingly concluded that the payment of £215 must have included interest on Sir Thomas Temple's original offer of £200 recorded a few days earlier (Orbison, p. 5 and document C. 11). Temple was, however, apparently induced to double his offer and duly received the stipulated first instalment of the repayment on 18 November 1613 (Orbison, C. 29). In the light of this information, Orbison's estimate of the total cost of the masque as having been £2,255 8*s*. 11*d*. (p. 7) should probably be raised by £200. Sir Thomas Temple

(1567–1636), a former member of Lincoln's Inn, had been created first Baronet of Stowe in 1611. Of the 'officers of this howse' who are mentioned in the document, Richard Baldwin was the Under-Treasurer (see above, 1. B), Francis Barker was the Butler, and Arthur Blunt was the Junior Butler until the following year (1614), when he rose to Second Butler.

2. Documents relating to Shirley's *The Triumph of Peace*

2. A. Masque roll, dated 16 October 1635, listing the taxations levied on individual members of the Middle Temple. It is now contained, with some other masque rolls, in a box mainly devoted to accounting papers of the period 1640–50. Endorsed: 'Masque T'inis Mich' Hil' Pasch' & Trinit' 1634–1635. 39^{li}–00s. & rec: on last yeares roll— 7^{li} in tot': 46.' This is the only taxation roll that has survived for either Chapman's or Shirley's masque.

16° Octobr': 1635.
M^{r}: ffrancklyn Treor':

Moneys rec: of m^{r}: Chafin gathered on the Masque roll this yeare, in the time of m^{r}: ffrancklyn Treor':

	li s d
In prim': 5.º Dec': 1634	10–00––0
Itm': 2.º Maij 1635	09–00––0
Itm': what he was areard on the last yeares role, excepting Pyne I	07–00––0
Itm': rec': more 24°. Iunij	10–00––0
Sum':	36–00––0

out of wch 36^{li} I pd: him 5^{li} allowed for his paines in Collecting the whole rolls.
So the whole rec': of him deducting as aforesd: is 31^{li}.

Also I rec': of

Southcott I: ———	1^{li}
Skippe R: ———	1.
[Morse T:] ———	0
Lydall R: ———	1.
Lewes H: ———	1.
Lister I: ———	1.
Bingham I: ———	1.
Dering A: ———	2.
Barton I: se': ———	1.
Boothby R: ———	1.
	10.
Sum': Totlis	46^{li}.

Orbison's documents, as well as several presented here, show that the Middle Temple was still paying off debts incurred for the production of masques many years after the event. On the same day that this roll was compiled Robert Thorpe, treasurer for Shirley's masque, recorded payments of more than £50 to various tradesmen (Orbison, S. 76). The collection of the taxation was in the hands of Francis Chafin, Second Butler of the Middle Temple, who had been fined 40*s*. for failing in some of his other duties with regard to the masque in January 1633/4 (*Minutes of Parliament*, ii. 814–15). The names of the members can all be found in Martin's index: John Pyne, John Southcott, Richard Skipp, Thomas Morse, Richard Lidall, Henry Lewis, John Lister, John Bingham, Antony Deering, John Barton senior, and Robert Boothby. Most of them were Utter Barristers, and all occupied chambers in the Inn at the time of the masque.

2. B. Copy of a warrant dated 21 November 1634, signatures omitted, asking Robert Thorpe, masque treasurer for the Middle Temple, to reimburse John Herne of Lincoln's Inn for £12 10*s*. to be paid by him to servants and tradesmen for taking part in the masque. Copied into the Treasurer's Receipt Book for 1637–8 (Fol. 16b), the copy dated 19 May 1638.

21 Novembris 1634:

Wee disire you forthwth to pay vnto Iohn Herne Esqr the Sum*me* of Twelue Pownd*es* tenn Shilling*es* being one fourth of the ffyfty pownd*es* by him to bee p^{d} over vnto serv^{tes} and oth^{rs} that attended the maske for their gratuity And this shalbee your warr^{t}./

To Rob*er*te Thorpe Esqr Trēr for the Maske for the middle Temple //.

Tho: Lane I pray you pay this warrant./

Rob*er*te Thorpe./

Receiued vpon this Bill the 19th of May Anno 1638 the Sume of nyne pownd*es* fower shilling*es*, the other monney is stayed & allowed m^{r} hide as disbursed for torches & silke stocking*es* } 09 04 00

Iohn Herne:/ Sume is 09^{li}04^{s}00^{d}:

The 'serv^{tes} and oth^{rs}' were employed to carry torches and to lead horses in the procession from the City to Whitehall that took place before the performance of the masque at court. Orbison S. 72 contains Herne's record of the payment, written in January 1634/5. The fact that he did not receive reimbursement from Thorpe until more than three years later — and then only for part of the amount — suggests a reason why the Middle Temple continued to collect money on masque rolls as late as 1639 (see below, 3. E. 15). Herne was the chief disburser of funds for Shirley's masque for the four Inns. Edward Hyde of the Middle Temple, later the first Earl of Clarendon, served on the masque committee and was one of those in charge of providing for the horses and riders in the procession. It is not clear why this document was copied into the Middle Temple annual Treasurer's Accounts, since the masque accounts were kept separately from the general accounts of each Inn. The first half of the entry only was printed by Hopwood,[4] whose version is reproduced in Orbison, pp. 38–9, n.2.

3. Documents Relating to Other Entertainments at the Middle Temple

Some of the miscellaneous papers kept in the 'Masques and Entertainments' box have also been published separately by Orbison and need not be further described here.[5] Among the others perhaps the most interesting is a receipt signed by John Dowland and two other musicians for a performance by a consort in the Middle Temple hall on Candlemas Day, 1612/13:

3. A. Musicians' receipt, dated 2 February 1612/13. Endorsed: 'Mr Dowland & his ffellowes for the consort ou' the feast day of the purificacion of our ladye anno 1612.'

Receaved by mee Iohn Dowland for my selfe and my fellowes Musitians, vpon candellmas Daye, j6i2 for the consorte p'formed before the Iudges and Reverent benchers, of the honourable Society, of the midle Temple by the handes of m^{r} Richard baldwine vnder Tresoror of the sayd midle Temple the some of five pound*es* sterlinge I saye by mee Receaved ———— 5lb

Io: dowlande
Lutanist to the King*es* ma^{tie}/
William Corkine
Richarde Goosey

In his signature Dowland flaunts his long-delayed appointment as a member of the King's Lutes, which was confirmed by royal warrant on 28 October 1612. The day after his concert at the Middle Temple, Dowland was paid £2 10*s.* as one of the lutanists in Chapman's masque.[6] William Corkine is known only from the two books of airs for viol accompaniment (unlike Dowland's airs, which were scored for lute) that he published in 1610 and 1612.[7] The name of Richard Goosey does not appear in any list of professional musicians for

[4] *A Calendar of the Middle Temple Records*, ed. Charles Henry Hopwood (London, 1903), p. 152.

[5] Orbison, 'Traces of Two Jacobean Dramatic Performances at the Middle Temple', *Yearbook of English Studies*, i (1971), 55–62.

[6] *Lincoln's Inn Black Books*, ed. William Pailey Baildon, 4 vols. ii (London, 1898), 156.

[7] *New Grove Dictionary of Music*, 'Corkine', and see Priska Frank, 'A new Dowland document', *Musical Times*, cxxiv, no. 1679 (January, 1983), 15–16.

the period. Unless he had a namesake, he was at this time an assistant steward of the Middle Temple and later leased a house belonging to the society just outside the main gate, where he worked as a scrivener (*Minutes of Parliament*, ii. 622).

In the same carton there is also a plea for payment by the house muscians during the year of Chapman's masque:

3. B. Musicians' Petition, undated, and endorsed three times: (1) (in the same hand as the petition) 'To the Righte Worshipfull the Maisters and benchers of the Middell Tempell'; (2) (in a second hand) 'The sume of xls ys to be allowed to them'; (3) (in a third hand) 'The musicions peticion where vppon theire was granted vnto them by order of p'liam't xls'.

We your poore orratoures the musissians intreates your worshipes for fower nightes paye, as namely the fifte of Nouember, the Satterdaye followinge, the satterdaye before the Princes funerall, and one Satterdaye after Chrismas

Though the document is undated, it is evident from internal evidence that the events it records took place during the period from All Saints' Day to Christmas week, 1612. Prince Henry's funeral was held on 7 December 1612. This dating is also confirmed by a related document, now to be found in a carton of miscellaneous accounting papers from the period 1610–29:

3. C. Musicians' receipt, dated 21 June 1613. Endorsed: 'The musicions Bill for xls when they gave their attendance abowt the tyme when the Prince dyed'.

Iune the xxjth Anno 1613
Receved of m^{r} bawldin forty shillings for foare nyghts sarvis of the mvsissions
by me Iohn Papson

Though the two documents are in different hands, B being a scribal copy and C being a holograph, it seems likely that the author of both was John Papson, who was the leader of the Middle Temple's house consort at the time, as well as a member of the City minstrels' guild. The payment of 40*s*. to the Christmas musicians had been authorized by parliament on 27 November 1612 (*Minutes of Parliament*, ii. 557). The six-month delay in paying them may have been because of a shortage of cash due to the expenses for Chapman's masque. As the Christmas musicians were normally paid in February of the following year, B may be roughly dated as having been written in February or March of 1613. The endorsement on C is in the hand of Richard Baldwin, the Under-Treasurer.

3. D. Musicians' receipt, dated 15 February 1618/19, signed by Andrew Tye, with an additional receipt by Henry Field, violinist, written on the same sheet. No endorsements.

ffebruarie the xvth 1618

Receved of mr Richard Balwin for the vse of my selfe and the reste of the mvscions of this howse the some of fortie shillinges in full discharge of the mvsicke for this yeare I saye Rec } xls

Andrew Tye

Reseved by henery feld for all holland daye for playing vppon the treble viallyn of mr balwin 4s

Nothing further is known of Andrew Tye. In 1619 Henry Field was appointed a member of the London Waits as a treble violinist, and he later played for both the Inner Temple house musicians and the band of the King's Men at the Globe and the Blackfriars.[8]

3. E. Bound Treasurer's Accounts, 1614–15, 1637–42

This completes the inventory of the documents in the carton labelled 'Masques and Entertainments' that were not published by Orbison. It remains to indicate the contents of the further six cartons of miscellaneous administrative and financial documents that survive for the period 1600–50, in so far as they concern dramatic and musical events at the Middle Temple. Three of these cartons, containing, respectively, Treasurer's Papers from 1624–5, Treasurer's Papers from 1630–9, and 'Seventeenth-century Correspondence', mainly after 1642, reveal no documents of interest. The remaining three, however, contain a number of relevant documents, of which the most important are six sets of bound Treasurer's Accounts. Some of these were seen by C. H. Hopwood at the turn of the last century,[9] but went missing until a librarian rediscovered them after enquiries in November 1989. Others were unknown to Hopwood and were turned up by an archivist in the 1970s. The extant volumes may be summarized as follows:

(1) Treasurer's Disbursement Book, 1614–15. 34 leaves, bound in paper. Kept in a carton labelled 'Treasurer's Papers, 1612–1617'.

(2) Treasurer's Receipt Books, 1637–8 and 1638–9. 91 leaves, bound in vellum. Separately shelved, no shelf-mark. (The receipts for the two years are bound together into a single volume.)

(3) Treasurer's Receipt Book, 1639–40. xii + 48 + vii leaves, bound in vellum. Separately shelved, no shelf-mark.

[8] Walter L. Woodfill, *Musicians in English Society from Elizabeth to Charles I* (Princeton, 1953), p. 249.

[9] The accounts known to Hopwood were the single volumes for the years 1637–40, and two of the four surviving volumes for the years 1640–2.

(4) Treasurer's Disbursement Book, 1640–1. 9 leaves, stitched at top. Kept in a carton labelled 'Accounting Papers, 1640–50'.

(5) Treasurer's Receipt Book, 1640–1. v + 23 + xvii leaves, bound in vellum. Kept in the same carton as (4).

(6) Treasurer's Disbursement Book, 1641–2. 7 leaves, stitched at top. Dated 8 May 1646, accompanied by a report by the accountant appointed to prepare the accounts for 1641–2 due to the death of the Treasurer while in office. Kept in the same carton as (4).

(7) Treasurer's Receipt Book, 1641–2. iv + 74 + iv leaves. Bound in vellum. The volume also contains the receipts for the years 1642–6.

The Treasurer's Disbursements for 1614–15 include payments to John Papson for the musicians' annual salary, and to the Steward for the musicians at Christmas:

3. E. 1. Undated. (Fol. 17a)

To Papson one of the musicions that *s*erveth the howse for his & his ffellowes yearely fee, payde in hillary Tearme 1614 xls

3. E. 2. Undated. (Fol. 27a)

payde vnto m^r Steward for the allowance of \ye/ Christmas Comons vidlt towarde*s* the musick xls

For John Papson see above, 3. C. As that document shows, at the Middle Temple the musicians were paid separately for their performances during the Christmas season in addition to receiving an annual salary in February. This was also the case at the other Inns of Court, since by custom the expenses for commons during the Christmas vacation were the responsibility of the Steward rather than of the Treasurer.

The receipts for 1637–8 and 1638–9 show payments for both musical and dramatic entertainments. In 1637–8 plays were performed in the Hall on All Saints' Day and Candlemas by the Queen's Men, who occupied the nearby Salisbury Court theatre and whose leader was the actor Anthony Turner:

3. E. 3. 4 November 1637. (Fol. 4a–b)

A plaie

Rec' of Iohn Baylie of the middle Temple Esqr the Sum*m*e of Tenn pownde*s* the 4th of November 1637: by the hande*s* of Thomas Lane I say rec' by mee Antho: Turnor in the behalfe of his fellowes the Queenes serv^tes att Salisbury Courte: } li s d 10 – 00 – 00

. . .

A dozen of longe staves for Barristers ————————	00 – 07 – 00
A doz' of Truncheons and Portage ————————	00 – 05 – 06
for three watchmen that kept the doore ————————	00 – 03 – 00

rec' this Sum*me* 00 – 15^{s} – 06^{d}.
Tho: Lane./

3. E. 4. 2, 7 February 1638. (Fols. 10a, 11a)

2° ffebr' 1637

A doz' of long staues	00	07	00
A doz' of Truncheons and portage	00	05	06
for 3 watchmen that kept the doore	00	03	00

rec'd this sume. Sume is $\overset{li}{00}$ – 15 – 06
Tho Lane./

. . .

7 ffebr' 1637.

Rec' of Iohn Bayliffe of the middle Temple London Esqr for a play there acted on the feast day of the Purificacion last past the Sume of Ten pownd*es* I say rec' for the said play ———————— } 10 00 00

Antho: Turner./ Sume is $\overset{li}{10}$ – 00 – 00

As usual in these accounts, neither play is named, and the only other expenses recorded in connection with them are for the security precautions of staves, truncheons, and door-keepers. Turner received his money from John Bayliffe, Under-Treasurer, through the latter's servant, Thomas Lane, who had also handled all the Middle Temple's payments for Shirley's masque.

In the same year payments were also made for both the annual salary of the musicians and for their services during the Christmas Revels.

3. E. 5. 15 February 1638. (Fol. 11b)

15 ffebr' 1637:

Rec' of the Society of the middle Temple o^{r} Salary or wages for the yeare past the Sum*me* of ffifty three Shilling*es* and fower pence I say rec' as aforesaid } 02 13 04
Thomas Hunter./

Sume is $\overset{li}{02}$ – 13 – 04

3. E. 6. 28 February 1638. (Fol. 14a)

Allowance to the Steward for
Christmas 1637:

	li	s	d
Inprimis towardes Comons for the Officers ————	03	00	00
for the Salary of the Musitians ————	02	00	00
for A loade of Coales ————	01	18	00

pd the Steward this Bill
the 28th of ffebr' 1637./ Summe is li 06 – 18 – 00
Rec' Will Lane:/

Although the word 'musicians' does not occur in E. 5, there can be little doubt that it is a musicians' receipt, as the wording, the amount, and the time of payment are all identical to those found in other musicians' receipts in the Middle Temple accounts (see E. 9 and E. 11, below). Thomas Hunter, however, does not appear elsewhere in the accounts, nor does his name appear in any other list of musicians for the period. William Lane had been the Butler of the society since 1619 and became Steward in 1634. He may have been related to Thomas Lane, servant to the Under-Treasurer (see above, E. 3). They should not be confused with Richard Lane, Master of the Bench, who was the Middle Temple Treasurer in 1637–8.

In 1638–9 there are further payments for plays on All Saints' Day and Candlemas:

3. E. 7. 7 November 1638. (Fol. 45a)

7 Novembr' 1638
Rec' from the Society of the Midd Temple London for a play acted before them there on the ffeast day of all St last past [I] the Summe of Tenn Powndes I say rec' by th'appoyntmt of Iohn Bayliffe ———— } 10l 00 00

By me Antho: Turner./
Summe is li 10 – s 00 – d 00

---

ffor the play on all St day 1638:

long staves a doz' ————	00	07	00
Truncheons a doz' & Portage ————	00	05	06
for 3 watchmen to keepe the dore ————	00	03	00

7o Nov: 1638 Sume is l 00 – s 15 – d 06
Tho Lane./

3. E. 8. 5 February 1639. (Fol. 57a–b)

5 ffebruar' 1638
Rec' of Iohn Bayliffe of the middle Temple Lond Esqr for a play acted there on the ffeast day of the purificac*io*n last the Sume of Tenn Pownd*es* I say rec' } 10 00 00

Sume is 10 – 00 – 00

---

		l	s	d
for the play on Candlemas day last.//				
Long staves a doz'		00	07	00
Truncheons a doz'		00	05	06
watchmen to keepe the dore		00	03	00

Sume is 00^{l} – 15^{s} – 6^{d}

Although the scribe has neglected to copy the signature from the Candlemas receipt, it was presumably again that of Anthony Turner, acting on behalf of the Queen's Men as he had at All Saints' and at both feasts in the previous year. The 1638–9 accounts do not record a payment to the musicians at Christmas, but their annual wages were paid as usual:

3. E. 9. 16 February 1638/9. (Fol. 59a)

16 ffebruar' 1638
Rec' the wag*es* & allowance of the Musitians from the Society of the Middle Temple London for the yeare last past the Sum*me* of fower mark*es* I say rec' by the hand*es* of Tho: Lane } 02 13 04:

Jeffery Collins:

Sume is 02^{li} – 13 – 04

Jeffrey Collins, a theatre-musician from the Cockpit in Drury Lane, who had also been a member of the Globe and Blackfriars band, here makes the first of several appearances in the Middle Temple accounts as leader of the house musicians (see below, E. 11 and E. 20).[10]

The 1639–40 accounts again record both musical and dramatic entertainment:

[10] Gerald Eades Bentley, *The Jacobean and Caroline Stage*, 7 vols. (Oxford, 1941–68), ii. 409.

3. E. 10. 9 November 1639. (Fol. 4a)

9º Nov: 1639

Receiued for a Playe Acted att the Middle Temple by the Players att the Cock Pitt on the feast-day of All S*tes* last past the Sum*me* of Ten Pownd*es* I say rec' by the hand*es* of Tho: Lane } 10 00 00

Will Beeston./ Sume is li 10 – 00 – 00

3. E. 11. 14 February 1639/40. (Fol. 15a)

14 ffebr' 1639

Rec' then from the Society of the Midd Temple the yearely penc*io*n or allowance of the said Society vnto the Musitians for their paynes & attendance the yeare past I say rec' as aforesd by the hand*es* of Tho: Lane } 02 13 04

Jeffery Collins./ l 02 – s 13 – d 04

Will Beeston was the proprietor of a company of actors known as 'Beeston's Boys' who played at the Cockpit (or Phoenix) Theatre in Drury Lane. Like the company from Salisbury Court that had performed at the Middle Temple in the two previous years, they were perhaps chosen for their proximity as much as for their quality, since both theatres were within an easy walk from the Temple. As Jeffrey Collins played in Beeston's band, it is possible that his fellow theatre-musicians made up the consort that played throughout the year for the Middle Temple revels.

The accounts for 1640–1 survive in two parts, a Disbursement Book and a Receipt Book, which constituted the original form of all the Treasurer's accounts. The Disbursement Book contains two theatrical references.

3. E. 12. Undated. (Fol. 2a)

It' arerag*es* collected on the Masque rol in all the 4 termes 10 – 00 – 00

3. E. 13. 2 February 1640/1. (Fol. 7a)

It' for a plaie p*re*sented in the hall on the feast of purification & for staues & watchmen vppon 2 seu*er*all bils entred li 10 – s 15 – 06 d

The Receipt Book for the same year gives further information about the Candlemas play, though not about the masque roll.

3. E. 14. 5 February 1640/1. (Page 5)

5.º ffebruar': 1640.

Rec' : then for a Playe, acted by vs &c : on Candlemas Day
last past, before the society of the Midd : Temple, the
sum*me* of Ten pound*es*, wee say rec' : in full for the sd :
Playe 10 00 00

Robert Axon
Iohn Lacy

3. E. 15. 2 February 1640/1. (Page 43)

Longe staves a doz : Truncheons a doz :' & portage, on Candlemas day last	00	12	06
ffor watch-men at the dore	00	03	00

Both Robert Axen and John Lacy were actors with Beeston's company,[11] and so we may conclude that it was again the Cockpit players who entertained the lawyers in this year. It is unclear why the society should have been collecting money on a masque roll in 1640/1, since the last known masque performed at the Middle Temple was Davenant's *The Triumphs of the Prince d'Amour* in February 1635/6. The word 'arerag*es*', however, may be explained by the entry in the Minutes of Parliament for 24 May 1639, which reveals that more than two years after that entertainment there was still 'an ancient areare of 12^{li} 8^{s} due & oweing [to the Steward] ever since the Christmas of the prince *d'amour*, an*no* 1635. conc*er*ning w^{ch}: no order as yet hath beene taken for paym^{t}:'.[12] Steps were taken to pay off this debt, but they were not immediately sufficient. A single paper fragment, now to be found in the same carton that contains the 1634–5 masque roll (see above, 2. A) shows that the attempt to raise money in 1638–9 yielded a sum of only £2.

3. E. 16. Dated 'T'inis Mich' Hil' Pasc' & Trinit' 1638. 1639.'

collect' vpon ye Masque roll

Holles G	1
Michell I	1
Sum'	2

p' me Walt'rum Copley

Walter Copley was the Butler of the Middle Temple from 1628 to 1648. Gervase Holles and John Michell were both Utter Barristers. Since Michell surrendered his chambers on 29 January 1638/9,[13] the collection was presumably made sometime before then. The collection of £10 'arerag*es*' on the 1640–1 masque

[11] Ibid., ii. 353, 495–6.

[12] Middle Temple Archives, 'Parliament Book 1626–1658', 210. Martin's transcription of this entry omits the sum of the debt (*Minutes of Parliament*, ii. 880).

[13] *Minutes of Parliament*, ii. 878.

roll (E. 12) was thus apparently intended to be added to the two pounds collected in 1638–9 to pay off the deficit on *The Prince d'Amour* in 1635–6.

The 1640–1 Receipt Book, though curiously enough not the Disbursement Book, also records the payment of the musicians' annual wages.

3. E. 17. 22 February 1640/1. (Page 6)

> Rec: from the hand*es* of Tho: Lane
> for the ffee this yeare last past for the Musitians, the
> sum*me* of foure Mark*es* I say &c. 02 13 04
> By me Iohn Gamble

Gamble's theatrical affiliation at this time, if any, is not known. He may have been a colleague of Jeffrey Collins at the Cockpit, since Collins again collected the musicians' wages in the following year. He is known to have played the violin and cornet, and eventually became a member of the King's Musick under Charles II.[14]

The 1641–2 accounts also survive in the form of separate Disbursement and Receipt Books. In the Disbursement Book yet another collection was made for masques.

3. E. 18. Undated. (Fol. 8a)

> It' collected arerag on the Masque rol 02 – 00 – 00

There is nothing further to indicate why such a collection should have been made in this year, unless part of the previously collected masque money had been diverted to pay the debt of more than £264 that had been run up for Christmas commons in 1638–9, a sum which the Steward begged the Masters of the Bench to raise 'least those to whome it is oweing should be vtterly vndone'.[15] A mere two pounds, however, could hardly have made much of a dent in whatever was left of this debt in 1642. There were no further collections of masque money after this year.

This was also the last year in which the Middle Temple made an annual payment to a consort of house musicians. The Disbursement Book shows the following:

3. E. 19. Undated. (Fol. 23a)

> It', p'd the Musitians their yeares wag*es* or salarie, li s d
> ending in ffebr'. 1641 02 – 13 – 04

[14] Andrew Ashbee, *Records of English Court Music, Volume I (1660–1685)* and *Volume II (1685–1714)* (Snodland, Kent, 1986–7), *passim.*

[15] 'Parliament Book 1626–1658', 210.

The Receipt Book for the same year gives further particulars:

3. E. 20. 18 February 1641–2. (Page 17)

18.o ffebr:' 1641.

> Rec:' then or: salary or yearly wages from the society
> of the Midd Temple for or selues & the rest of o:r
> company the Musitians the sum*me* of 4: m'k*es* I say rec:'
> as aforesd: by the hand*es* of Tho: Lane 02–13–04
> Jeffery Collins[16]

Jeffrey Collins, who led the Middle Temple musicians in 1638–9 and 1639–40, was evidently the last person to hold this position, as no further payments of the sort are recorded during the 1640s. The reason for this is given in the following order by the Masters of the Bench in their session of 25 November, 1642.

3. E. 21. Parliament Book, 1626–58. (Page 273)

> It is ordered by the M:rs of the Bench of this p*ar*liamt that Com*m*ons maye be kept in the house, & so continue throughout the Christmas following, vnto the next Hilarie terme wthout any Musicke, Gaming or any publique noise or shewe, wherby companies may be drawne into the house; And this in respect of the danger & troublesomnes of the times.[17]

[16] This document was calendared by Hopwood, who, however, omitted the reference to musicians (*Records*, 155).

[17] Also printed in *Minutes of Parliament*, ii. 928.

APPENDIX 1

CHRONOLOGICAL SUMMARY OF DOCUMENTS

Years given are accounting years, beginning on All Saints' Day. References are to document numbers used in the text.

1612–13

25 January. Loan Receipt. (1. A.)
£50 from Francis Moore for Chapman's masque.

2 February. Musicians' Receipt. (3. A.)
£5 to John Dowland and others for a 'consorte'.

21 February. Admission Certificate. (1. C.)
Inigo Jones admitted to the Middle Temple by special election.

24 February. Loan Receipt. (1. B.)
Loan of £66 13*s.* 4*d.* from Sir Henry Mountagu for Chapman's masque.

?February. Musicians' Petition. (3. B.)
Request for payment for Christmas Revels.

11 June. Parliament Minutes. (1. D. 1)
Taxation of members for Chapman's masque.

16 June. Parliament Minutes. (1. D. 2.)
Revised taxation of members for Chapman's masque.

18 June. Parliament Minutes. (1. D. 3.)
Loan of £50 from Edward Powell for Chapman's masque.

18 June. Parliament Minutes. (1. D. 4.)
Loan of £400 from Sir Thomas Temple for Chapman's masque.

21 June. Musicians' Receipt. (3. C.)
£2 to John Papson for the Christmas musicians.

1614–15

Hilary Term. Treasurer's Disbursement Book. (3. E. 1.)
£2 to John Papson for the annual salary of the musicians.

Hilary Term. Treasurer's Disbursement Book. (3. E. 2)
£2 to the Christmas musicians.

1618–19

15 February. Musicians' Receipt. (3. D.)
£2 to Andrew Tye for the annual salary of the musicians.
4*s.* to Henry Field, violinist, for playing on All Saints' Day.

1634–5

16 October. Masque Roll. (2. A.)
£82 collected for Shirley's masque.

1637–8

4 November. Treasurer's Receipt Book. (3. E. 3.)
£10 to Anthony Turner for a play by the Queen's Men.

2 February. Treasurer's Receipt Book. (3. E. 4.)
£10 to Anthony Turner for a play on Candlemas.

15 February. Treasurer's Receipt Book. (3. E. 5.)
£2 13*s.* 4*d.* to Thomas Hunter for the annual salary [of the musicians].

28 February. Treasurer's Receipt Book. (3. E. 6.)
£2 to the Christmas musicians.

19 May. Treasurer's Account Book. (2. B.)
£12 10*s.* to servants and tradesmen who took part in [the procession for] Shirley's masque.

1638–9

Undated. Masque Roll. (3. E. 16.)
£2 collected.

7 November. Treasurer's Receipt Book. (3. E. 7.)
£10 to Anthony Turner for a play on All Saints' Day.

5 February. Treasurer's Receipt Book. (3. E. 8.)
£10 for a play on Candlemas.

16 February. Treasurer's Receipt Book. (3. E. 9.)
£2 13*s.* 4*d.* to Jeffrey Collins for the annual salary of the musicians.

1639–40

9 November. Treasurer's Receipt Book. (3. E. 10.)
£10 to William Beeston for a play on All Saints' Day by the Cockpit players.

14 February. Treasurer's Receipt Book. (3. E. 11.)
£2 13*s.* 4*d.* to Jeffrey Collins for the annual salary of the musicians.

1640–1

Undated. Treasurer's Disbursement Book. (3. E. 12.)
£10 collected on the Masque Roll.

2 February. Treasurer's Disbursement Book. (3. E. 13.)
£10 15*s.* 6*d.* for a play on Candlemas.

5 February. Treasurer's Receipt Book. (3. E. 14–15.)
£10 to Robert Axen and John Lacy for a play on Candlemas.

22 February. Treasurer's Receipt Book. (3. E. 17.)
£2 13*s*. 4*d*. to John Gamble for the annual salary of the musicians.

1641–2

Undated. Treasurer's Disbursement Book. (3. E. 18.)
£2 collected on the Masque Roll.

Undated. Treasurer's Disbursement Book. (3. E. 19.)
£2 13*s*. 4*d*. for the annual salary of the musicians.

18 February. Treasurer's Receipt Book. (3. E. 20.)
£2 13*s*. 4*d*. to Jeffrey Collins for the annual salary of the musicians.

1642–3

25 November. Parliament Minutes. (3. E. 21.)
Bench order forbidding music and public shows at Christmas.

APPENDIX 2

ACTORS MENTIONED IN THE DOCUMENTS

Axen, Robert (3. E. 14.)
Beeston, William (3. E. 10.)
Lacy, John (3. E. 14.)
Turner, Anthony (3. E. 3–4., 7.)

APPENDIX 3

MUSICIANS MENTIONED IN THE DOCUMENTS

Collins, Jeffrey (3. E. 9., 11., 20.)
Corkine, William (3. A.)
Dowland, John (3. A.)
Field, Henry (3. D.)
Gamble, John (3. E. 17.)
Goosey, Richard (3. A.)
Hunter, Thomas (3. E. 5.)
Papson, John (3. C., 3. E. 1.)
Tye, Andrew (3. D.)

FIREWORKS FOR QUEEN ELIZABETH

These proposals for a fireworks display for Queen Elizabeth have been edited by C. E. McGee from documents in the Pepys Library, Magdalene College, Cambridge, and checked by the General Editor.

November 1992 N. W. BAWCUTT

The Pepys Library at Magdalene College, Cambridge, holds a remarkable example of one of the spectacular kinds of show prepared for the entertainment of Elizabeth I. PL 2503, Papers of State, vol. II, fols. 607, 609 is an undated letter to the Earl of Leicester from Sir Henry Killigrew, who encloses a proposal for three nights of fireworks to be made by an anonymous Italian craftsman. Killigrew's letter suggests what costs a minor courtier like himself might bear to entertain the Queen—seven pounds for an impressive, unusual fountain—and what far greater costs a lord of Leicester's dignity would consider—fifty pounds and two months' work for a splendid display of fireworks.

Most obviously, Killigrew confirms, as Iago complained, that 'Preferment goes by letter and affection' (*Othello*, I. i. 36). Killigrew's letter illustrates how an artist might find his way into the profitable service of Leicester. Killigrew knew Leicester had a turn to be served and he knew of a capable artist eager to serve. His letter is a strong recommendation of the fireworks master; Killigrew mentions all those qualities which make the Italian a good risk: his proven skill, his ability to give realistic estimates, his economical suggestion for managing his advances, and his readiness to have his work inspected as it proceeds. What the letter shows, all in all, is a case of hiring by the Elizabethan grapevine.

Killigrew wrote in English in his own secretary hand on paper marked with a hand and flower. The enclosed proposal, written in Italian, is on a different paper; it is the same size as Killigrew's letter, but has slightly different laid lines. Presumably the Italian himself set it down in its neat format and handsome italic script. The beauty of the description of the proposed fireworks and the care with which it was prepared—marred by the omission of a mention of part of the second night's show—make it an advertisement promising careful, reliable service. Writing in Italian may have increased the force of his appeal for work, for it quietly flatters Leicester for his knowledge of foreign languages.

E. K. Chambers mentioned this fireworks device in *The Elizabethan Stage* (Oxford, 1923), i. 139 n.2, listing the various locations of each night's display and noting that the proposal might have been related either to the entertainment at Warwick in 1572 or to that at Kenilworth in 1575. The Historical Manuscripts Commission printed an accurate transcription of Killigrew's letter and an incomplete free translation of the proposal he enclosed in the *Report on the Pepys MSS. preserved at Magdalene College, Cambridge* (London, 1911), p. 178, and suggested 1575 as the date of the letter because the princely pleasures at Kenilworth Castle that year included some fireworks. R. J. P. Kuin, in his edition of Robert Langham's *Letter* (Leiden, 1983), includes a transcription of Killigrew's letter (pp. 6–7) and the proposed fireworks (Appendix C) on the assumption that they were for the Kenilworth entertainment of 1575.

The date of the fireworks device remains uncertain, however. Leicester was responsible for an entertainment at Warwick and two at Kenilworth that had fireworks. The Italian could have designed his display with the landscape of either place in mind, for both had castles with courtyards and nearby rivers and meadows. However, neither the fireworks in the mock siege at Warwick nor those in the shows at Kenilworth resemble the fireworks described in this proposal. If Leicester employed him in 1575, he purchased the Italian's fire-power, not his fireworks, for Robert Langham notes how high, how loud, and how brilliant the explosions were but makes no mention of figures in fire such as birds, dogs, cats, or a dragon. Similarly, apart from the dragon, 'flieing, casting out huge flames and squibes',[1] with which the show at Warwick came to a climax, the fireworks there had little in common with those proposed by the Italian. The most intricate device used at Kenilworth—fireballs that burned in water (symbols, says Gascoigne, of Leicester's unquenchable desire)—appeared in the entertainment at Warwick, but were not proposed by the Italian fireworks master. The Longleat Dudley Papers, vol. iii, fols. 55–6, record payments for materials needed for an exhibition of fireworks at Kenilworth in 1572; unfortunately the itemization of necessary supplies provides no clue to the character of the shows.

Killigrew's reference to his travels with Lord Hunsdon suggests a different, more plausible date. Of course it remains a conjecture, for Killigrew's hope that he may 'goe ouer my self this iornay wth my L. of honsden' (ll. 20–1) leaves his destination vague and the trip merely possible. In the Spring of 1564, Killigrew accompanied Hunsdon to Lyons for the ratification of the Treaty of Troyes. If France be the intended destination, then 1564 is likely to be the year in which he recommended this Italian and his fireworks device. In the early 1560s Killigrew increasingly sought the patronage of Leicester, and recommending a reliable artist may have been one way of ingratiating himself. Their association was well established by 1566, when Mary, Queen of Scots, referred publicly to Leicester as Killigrew's 'great frynd'.[2] At the same time, Leicester was Elizabeth's favourite suitor, a suitor eager to entertain her and, consequently, in need of capable artists and craftsmen. He might well have had in mind some long-term plans to entertain the Queen at his newly acquired estate of Kenilworth. The Queen awarded him the estate in 1563, and Leicester

[1] J. Nichols, *The Progresses and Public Processions of Queen Elizabeth*, 3 vols. (London, 1823), i. 320.

[2] PRO, SP 12/40/135–6.

renovated it in the next years to prepare for her first visit there in 1565. Killigrew also wrote to Leicester on 10 July 1564, to inform him that Lord Hunsdon had given him permission to remain in France to conduct some personal business with a M. de Beauvais.[3] This personal business would explain why Killigrew hoped to be able 'to goe ouer' with Hunsdon in the first place. The proposed fireworks device and Killigrew's letter of recommendation of the artist who could produce that device seem natural products of the relationship of Killigrew and Leicester in the early 1560s.

This document is printed by permission of the Master and Fellows, Magdalene College, Cambridge, and with special thanks to the former librarian there, Mr R. C. Latham, for his information about the manuscript, to Professor John C. Meagher for correcting the transcription, and to Dr. Vera Golini for help with the translation.

[3] *Report on the Pepys MSS.*, p. 29.

The man that desired me to present this inclosed vnto yor L, wold gladly know yor plesure therin for yt wyll ask too monethes work. yf therfor yō lyk his devyse, yt may plese yō, to tak ordre wth m^{r} dudley or som other, for the fornisching of hem wth mony. by his account the chargis wyll draw to .l. l̄i. wch som he desiryth not to haue in his owne handes, but that he may receve yt by iiij or .v. l̄i at a tyme, and wold gladly also that som by yor L. appoyntment may se how he dothe Imploy the same. the man is honest, and I think wyll sarve yor torne very well and far better in dead then in wordes. the vij l̄i wch he had of me, is Imployed abowt a fountayne wch he myndyth to present vnto the quynes m^{te} a singuler peace of worke (as I am Informed) wherof the lyk was neuer seane in these partis. I besiche yor L. to lett hem know yor plesure by my brother or som other, for that I think [myself] to goe ouer my self this iornay wth my L. of honsden, yf he obtayne leaue for me as I trust he wyll.

Yo^{r} good Lordshipes
most humble to comand

HKyllygre'.

2 *wold*] *w* altered 5 *dudley*] Mr John Dudley, servant to Leicester; see *Calendar of State Papers, Domestic Series, Elizabeth, Addenda 1566–1579* (London, 1871), 2–4 19 *my brother*] William, his younger brother, for his elders (John, Thomas, and Peter) were pirates 21 *self*] added in the left margin *honsden*] Henry Carey, lord of the manor of Sevenoaks in Kent and of lands in Yorkshire and Derbyshire, who entertained the Queen at Hunsdon House in 1571 (J. Nichols, *Progresses of Queen Elizabeth*, i. 282–3. As Governor of Berwick, Carey was involved in the Queen's affairs in the North in the early 1570s, including the execution of Thomas Percy and the disputes that arose from the claims of Mary Queen of Scots *HKyllygre'*] *H* and *K* conjoined and the signature underscored with a flourish

La prima sera ne'l prato.

Si faranno certi artificij doue si vedranno discorrere a torno certi serpenti di fuoco. Il che sara cosa molto piaceuole.

Item otto o dieci pignate con inuentioni di cose marauigliose & piaceuoli

Item de le aui uiue uolare atorno ne l'aria le quali getteranno fuoco da per tutto.

Item due cani & due gatti uiui li quali artificiosamente combattranno.

La seconda sera ne'l Cortile del palazzo.

Si uedrà un fonte dal quale scorrera vino acqua & fuoco sette o ott' hore continue. Qual fonte sara cosa degna di uedere per gli suoi marauigliosi artificij quali per essere tanti si lascia di scriuere.

Item tre ruote di fuoco mirabili & odorifere, & di diuersi colori.

La terza sera nel fiume.

Si uedra un dragone grande come un bue. quale volera due o tre uolte più alto che la torre di San Paolo, e stando si alto si consumera tutto di fuoco, & indi usciran subito da tutto'l corpo cani, e gatti & uccelli li quali uoleranno, & getteranno fuoco da per tutto che sarà cosa stupendissima.

28 *serpenti*] See John Babington, *Pyrotechnia, or a discourse of Artificiall Fire-Works* (London, 1635), p. 14, on how these were made 29 *pignate con inuentioni*] See Eberhard Fahler, *Feuerwerke des Barock* (Stuttgart, 1974), pp. 13, 25, where tubs of fire are shown in the courtyard of a castle 33 *combattranno*] For the machinery by which such fights were run, see Babington, *Pyrotechnia*, fig. 11, Fahler, pp. 53, 55, and 59, and Robert Norton, *The Gunner: the Making of Fire Works* (London, 1628), p. 155 and Tract. 3, ch. xxviii 35 *un fonte*] For devices blending fire and water, see Babington, *Pyrotechnia*, pp. 55–60 36 *continue*] *e* ends with flourish resembling large apostrophe 39 *Item . . . colori*] apparently omitted by mistake; interlined between 38 and 40. For wheels of fire, see Babington, *Pyrotechnia*, pp. 21–5, and Norton, *The Gunner*, p. 153. For the mixture of chemicals for various colours of fire, see Babington, *Pyrotechnia*, p. 11 (for red, white, and blue flame) and p. 14 (for gold and silver) 40 *fiume*] *e* ends with flourish resembling large apostrophe 41 *un dragone*] See Babington, *Pyrotechnia*, figs. 10 and 11, Fahler, *Feuerwerke des Barock*, p. 15, and Norton, *The Gunner*, ch. lxxvii 42 *volte*] *o* altered 44 *uoleranno*] [1]*o* altered 45 *sara*] [1]*a* altered

Vi sono molte altre cose in questi artificij le quali per la lor
difficultà non scriuo minutamente. Io le farò tutte
benissimo secondo il danaro che per le spese mi sarà mandato.

[The first evening in the meadow.

There will be constructed some works of art where serpents of fire will be seen spinning round, which will be very pleasant. Item eight or ten pots [of fire] with inventions of things marvellous and pleasing. Item some life-like birds to fly around in the air, who will throw out fire in all directions. Item two life-like dogs and two cats who will be skilfully made to fight.

The second evening in the courtyard of the palace.

A fountain will be seen from which wine, water, and fire will run for seven or eight hours continuously. This fountain will be a thing worth seeing for its marvellous effects, which are too numerous to describe. Item three wonderful wheels of fire, odoriferous and of diverse colours.

The third evening on the river.

A dragon as big as an ox will be seen, which will fly two or three times higher than the tower of St Paul's, and staying at that height it will be wholly consumed with fire, and then suddenly there will issue from the whole of its body dogs, cats, and birds which will fly about and throw out fire everywhere, which will be a most stupendous thing. There are many other things in these works of art which because of their complexity I do not describe in minute detail. I will make them all very well indeed in proportion to the money that will be sent me for expenses.]